VICTORIA 3
Complete Guide

Alex Wilkinson

Copyright © 2022 Alex Wilkinson

All rights reserved.

ISBN: 979-8-3604-3325-5

Victoria 3 Complete Guide

1	Victoria 3: Beginner Tips	8
2	Understanding Pops	9
3	Understanding Goods	10
4	Always Exploit Natural Resources	11
5	Think Carefully About Research	12
6	Don't Expand Too Quickly	13
7	Don't Be Afraid To Get Into Debt	14
8	Savescumming Works At War	16
9	VICTORIA 3 STARTER TIPS	17
10	VICTORIA 3 STARTER TIPS: USE THE MARKET QUICK MENU	18
11	VICTORIA 3 STARTER TIPS: CHECK YOUR RULER'S TRAITS AND IDEOLOGIES	19
12	VICTORIA 3 STARTER TIPS: COLLECT OBLIGATIONS	20
13	VICTORIA 3 STARTER TIPS: BUILD UP YOUR GOLD RESERVES	21
14	VICTORIA 3 STARTER TIPS: EXPLOIT YOUR NATURAL RESOURCES	22
15	VICTORIA 3 STARTER TIPS: UNPRODUCTIVE TRADE ROUTES CAN STILL BE USEFUL	23
16	VICTORIA 3 STARTER TIPS: UNIFY NATIONS TO EXPAND	24

17	VICTORIA 3 STARTER TIPS: DON'T RESEARCH TECHS YOU ARE GETTING TECH SPREAD FOR	25
18	VICTORIA 3 STARTER TIPS: FOCUS ON BUILDING BASIC GOODS FIRST	26
19	VICTORIA 3 STARTER TIPS: USE CONSUMPTION TAXES ON LUXURY GOODS	27
20	BEST COUNTRIES TO PLAY IN VICTORIA 3	28
21	BEST COUNTRIES TO PLAY IN VICTORIA 3: GREAT BRITAIN	29
22	BEST COUNTRIES TO PLAY IN VICTORIA 3: INDIAN TERRITORY	30
23	BEST COUNTRIES TO PLAY IN VICTORIA 3: EGYPT	31
24	BEST COUNTRIES TO PLAY IN VICTORIA 3: CAPE COLONY	32
25	BEST COUNTRIES TO PLAY IN VICTORIA 3: GREAT QING	33
26	THE WHOLE WORLD AT YOUR FINGERTIPS	34
27	HOW TO BE AN OMNISCIENT RULER	35
28	LINE GO UP SIMULATOR	37
29	THE GRANDEST OF STRATEGY	39
30	HOW TO INCREASE GDP IN VICTORIA 3	40

31	HOW TO INCREASE GDP IN VICTORIA 3	41
32	HOW TO INCREASE LEGITIMACY IN VICTORIA 3	46
33	HOW TO INCREASE LEGITIMACY IN VICTORIA 3	47
34	HOW TO INCREASE INFLUENCE IN VICTORIA 3	51
35	HOW TO INCREASE INFLUENCE IN VICTORIA 3	51
36	HOW TO INCREASE AUTHORITY IN VICTORIA 3	55
37	HOW TO INCREASE AUTHORITY IN VICTORIA 3	55
38	HOW TO INCREASE BUREAUCRACY IN VICTORIA 3	59
39	HOW TO INCREASE BUREAUCRACY IN VICTORIA 3	59
40	HOW TO INCREASE LITERACY IN VICTORIA 3	62
41	HOW TO INCREASE LITERACY IN VICTORIA 3	63
42	VICTORIA 3 TIMESPAN: WHAT IS IT?	67
43	VICTORIA 3 TIMESPAN	67
44	Victoria 3: How To Manage Your Construction	69

45	Types Of Building	70
46	Arable Land And Subsistence Farming	71
47	Urban Centers	72
48	Trade Centers	73
49	Government Administration	74
50	Infrastructure	75
51	Construction Sectors	76
52	When To Expand Buildings	77
53	Victoria 3: How To Raise And Lower The Price Of Goods	78
54	Buy Orders And Sell Orders	79
55	The Types Of Goods	81
56	The Importance Of Infrastructure	83
57	Victoria 3: How To Raise Your Standard Of Living	84
58	What Standard Of Living Affects	84
59	How To Raise Wages	86
60	How To Raise Welfare Spending	87
61	Victoria 3: How To Manage The Market	89
62	National Markets	90
63	Imports And Exports	91
64	Tariffs And Market Goods Policies	92

65	Importing To Supplement Production And Needs	94
66	Convoys	95
67	Victoria 3: Complete Guide To Warfare	96
68	Diplomatic Plays - AKA, How To Start A War	97
69	Escalation And Maneuvers	99
70	An Introduction To Warfare	101
71	Warfare Tips - AKA, How To Win A War	106
72	Victoria 3: How To Manage Your Interest Groups	108
73	Interest Groups And How They Form	109
74	Ideaologies And Traits	110
75	Approval	111
76	Clout	112
77	Bolstering And Suppression	114

VICTORIA 3: BEGINNER TIPS

Victoria 3 is an incredibly complex game - here are some tips to help you get your feet wet.

Victoria 3 is a complicated game. It's an incredibly deep, complex grand strategy game with a heavy focus on economic and political simulation. As a result, there are a ton of moving parts, and disruption to just a few can lead to catastrophic events.

It's for this reason that getting to grips with the basics and mastering them early is key to success. You'll have to get your head around your population, the obstacles you'll face when trying to pass new laws, and the complex market system if you want to soar - otherwise, you'll crash and burn your way into the twentieth century.

Understanding Pops

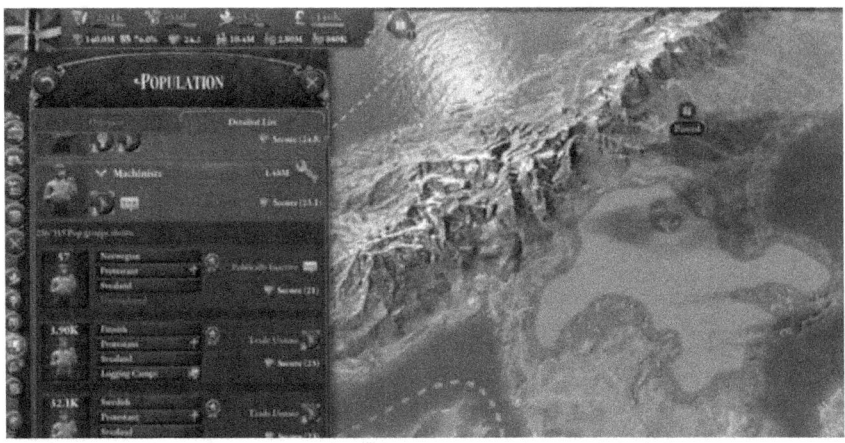

Understanding what Pops are is a very important part of getting to grips with Victoria 3. Pops is short for Population, and they are a stratified representation of the people who live in your country.

A Pop is categorized by their Profession, Culture, Religion, and Workplace, and each unique combination of these four variables is its own Pop. A Pop's size is simply the number of people who fit all four specific variables.

For example, Catholic Welsh Peasants who work on a Rye Farm would form one Pop, but Protestant Welsh Peasants who work on a Rye Farm would be a separate one.

Pops are your country's lifeblood - they staff your buildings, form your armies, and pay your taxes. They also vote or at least have the potential to be politically active. Depending on your laws, the goods you have access to, and your economic stability, individuals may end up joining

Interest Groups, with certain groups being more attractive to certain Pops. These are politically motivated movements that have varying levels of clout - they can be used to your advantage to pass laws that you're in favor of, or they can prove to be a hindrance, stopping you from doing what you like or even revolting at points.

Understanding Goods

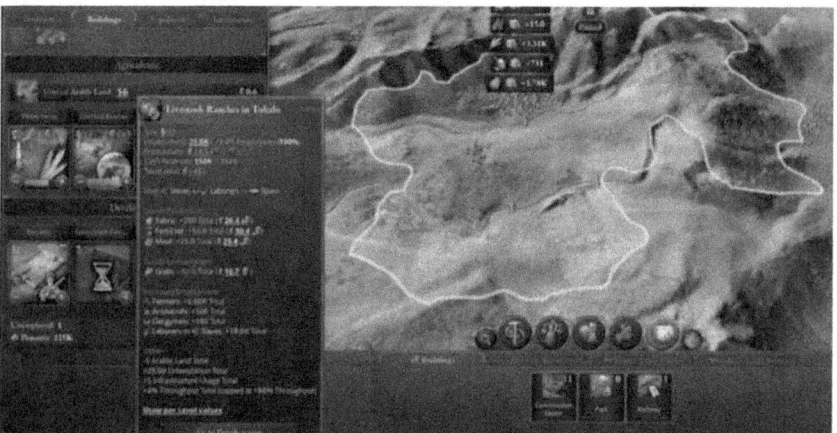

While Pops are the lifeblood of a nation, Goods keep the blood pumping. Goods have three general uses:

- Goods can be used in buildings to create or supply the production of different goods. For example, Iron is used to make Tools.

- Goods are used by Pops to satisfy their wants and needs. For example, Grain is used to feed people, and Liquor is used to satisfy a desire for substances.

- Goods can also be traded on the market for profit.

Keeping a healthy balance of Goods is important. If you don't have enough of a certain type of Goods, you may:

- Not be able to fully supply a building, meaning its output won't be sufficient enough to keep it in balance. This leads to lower wages, fewer employees, and knock-on effects in the supply chain.

- Not be able to supply Goods to Pops, lowering their standard of living. Pops with an insufficient standard of living may lead to turmoil.

- Not be able to secure profitable export trade routes.

Of course, oversupply can also be an issue - though less of an issue than undersupply. Having too much of a Good means its market price will plummet, which may lead to certain buildings not being able to turn a profit. This leads to a lower balance and lower wages.

Always Exploit Natural Resources

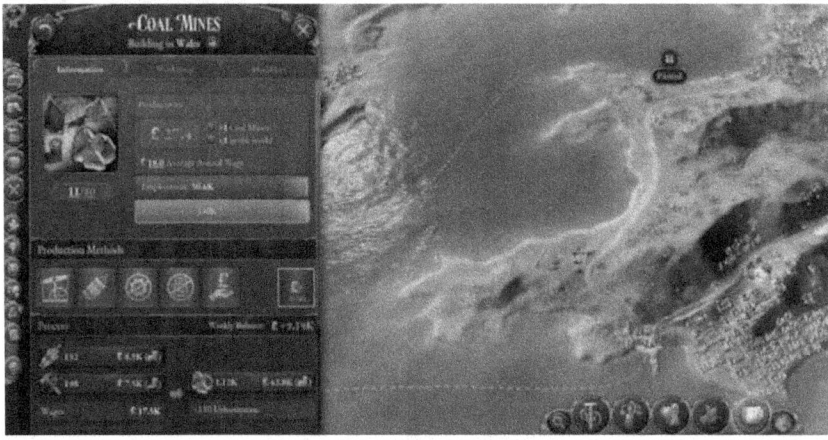

Your nation's natural resources are incredibly important - they will be crucial to your economy for much of the game if you go heavy on trade, and the more resources you have, the better you'll be able to supply your own luxury goods construction lines.

Even before you have a stable, self-maintaining economy, setting your resource buildings to upgrade automatically is a good way to ensure a steady income of raw goods. This is especially true of widely used goods, such as Wood, Coal, Iron, and Grain.

Be careful that you don't oversupply, though, as prices will plummet!

Pumping lots of money into raw goods production is a great way to earn prestige by becoming one of the world leaders in the good's production!

Think Carefully About Research

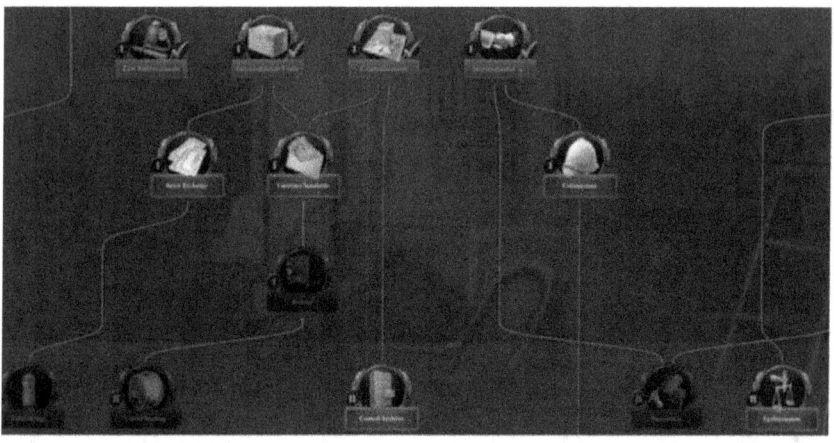

Unlocking new methods of production or goods to produce always feels great, but you'll want to hold off on the upgrades until you can comfortably sustain what they ask for.

Victoria 3's technology tree is very wide, and technology spread can be quite random depending on how other nations develop. This means the upgrades and unlocks you get might not be in the most logical order - it's easy to fall into a trap of just researching the techs that sound good, but you can be a lot more efficient if you just take the time to think things through.

Mid and late-game goods and upgrades require a large number of input goods to get off the ground. Even if you're a significant supplier of the raw materials of a good, you'll want to consider whether you'll be able to sustain the production levels that the potential upgrade asks for.

On the flip side, it's very possible to unlock certain goods before you even have a use for them. Stand-out examples of these are Electricity and Oil - don't construct Power Plants or Oil Rigs until you have other buildings and production methods that use the goods they produce - or know that they'd be great on the markets.

Don't Expand Too Quickly

Depending on your starting nation, you may be in a very good position to conquer your neighbors' lands quite easily, whether this is through powerful alliances or your own powerful army.

Be wary when considering this style of gameplay, though. A conquered state needs to be integrated as soon as you can, requiring a hefty chunk of Beaurocracy to do so, and it is very likely to be in turmoil for a number of years. This contributes to your level of radicals and can pose a significant threat.

To integrate a state, you need to head to its overview tab - which may not be something you look at lots! Try to make it a habit after every war.

Don't Be Afraid To Get Into Debt

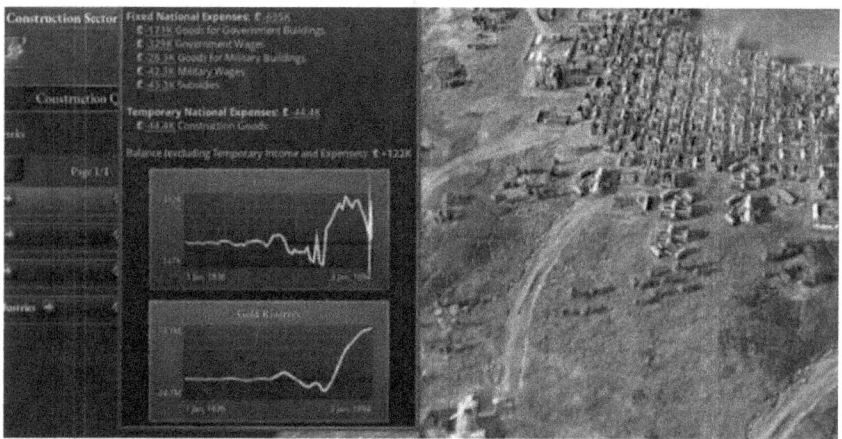

Victoria 3 isn't like other grand strategy games - it tries to simulate a realistic economy down to the ability to borrow capital from populations. As your balance grows, you'll stockpile it in the form of gold, which forms your treasury. Your gold reserves have maximums, though, so it's always a good idea to be spending money on construction or even costly trade routes if they get you the Goods you need to run your country.

You can increase your Gold reserve maximums with technology!

What this means is that you don't need to panic if you suddenly see your balance hitting the negatives and you start losing your accrued gold stores. You don't even need to panic if you hit zero! If you run out of gold and still have a negative weekly balance, you'll start borrowing from your Pops - this is called credit, and there's a certain amount you can take before getting into real trouble.

You can check your level of credit by hovering over the balance icon.

If your credit amount reaches the maximum and stays there for a while, you may have to declare bankruptcy. This will cause significantly severe negative effects for your nation, including wiping out your buildings' balances and increasing the overall level of radicalism. Try to avoid having to default as much as possible.

Tips for getting out of a negative balance in a pinch include:

- Freeing up some bureaucracy.
- Adding consumption taxes on some of your pricier goods.
- Increasing general taxes in the short term.
- Pausing construction.

Savescumming Works At War

If you're the type of person who doesn't mind doing so, savescumming is a good way to get the outcomes you desire during the early stages of a military campaign. If you save before starting a diplomatic play that would lead to a war, you can scope out the countries that are more likely to join either side and which ones can be swayed. In many cases, it will be possible to reload and sway key allies to your side before your foe snags them.

Alternatively, you can give up on the war before you ever start it if the odds are too mismatched in your enemy's favor.

This is not possible in Ironman mode, but remember that it is possible to get achievements without being in Ironman mode in Victoria 3!

It feels a little like cheating, but it's certainly one route to success.

VICTORIA 3 STARTER TIPS

You will need our Victoria 3 starter tips if you want to get ahead with your nation and climb the prestige ranks to become a Great Power. Victoria 3 is a complex grand strategy that focuses on economic management and political manoeuvring, allowing you to carve out your nation's future and make a name on the world stage. With a plethora of interconnecting systems and mechanics, you may need help taking advantage of them to prosper and grow, so check out our Victoria 3 starter tips.

VICTORIA 3 STARTER TIPS: USE THE MARKET QUICK MENU

You'll likely spend a lot of your time in Victoria 3 poring over the market screen, where you can take a look at the buy and sell orders of your goods, along with their balance and overall market price. It's a vital screen for analysing your economy and seeing what goods need more production, but by right-clicking any good it becomes even more useful. This will bring up a menu of actions you can take, such as establishing trade routes, setting market policies, expanding production, and expanding consumption, saving you from navigating to the various other menus to perform the same tasks.

VICTORIA 3 STARTER TIPS: CHECK YOUR RULER'S TRAITS AND IDEOLOGIES

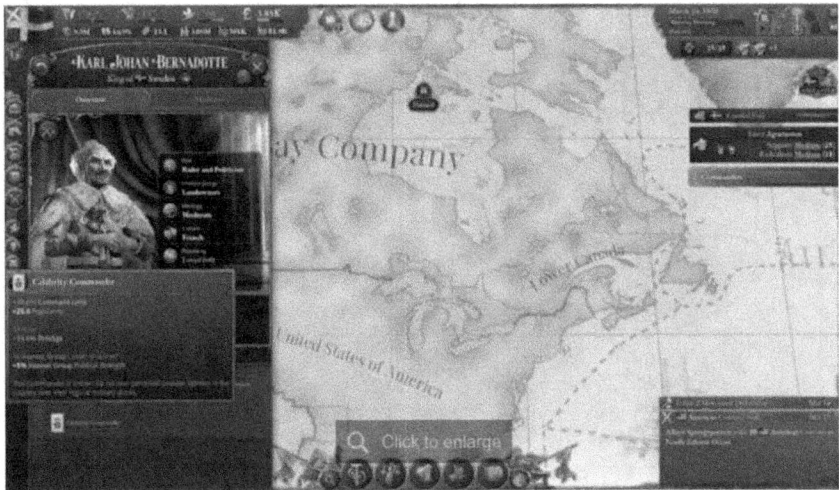

On the politics screen, you can take a look at the current ruler of your country, which can be a king, president, dictator, and much more, but each of them have their own unique traits and ideologies. These can give massively useful bonuses that are worth taking into account when making decisions about your country. For example, you could receive discounts to any decrees you use, or increased power projection. However, when you lose these benefits, they can have the effect of reducing your Capacities or prestige, so make sure you're ready for a new ruler to takeover.

VICTORIA 3 STARTER TIPS: COLLECT OBLIGATIONS

Obligations are incredibly powerful tools to be used for diplomacy, and you should try to get as many as you can from other countries during your game. They can be gained as a result of agreements made during Diplomatic Plays or by buying out a country's loans. When a country has an obligation toward you, they cannot start Diplomatic Plays against you, or join sides opposing you, and you can use the obligation to sway them toward your side during a Diplomatic Play, giving you an advantage or gaining you an ally if war breaks out.

VICTORIA 3 STARTER TIPS: BUILD UP YOUR GOLD RESERVES

Click to enlarge

Like any Paradox grand strategy game, money is incredibly important, and it's used for a variety of functions, such as paying wages or constructing buildings. Whenever you have a positive income, you will begin to build up gold reserves as a safety net, and while letting this fill can have adverse effects on your economy, it's important to have a surplus of money when making lots of buildings. Constructing buildings costs a lot of money, and will often bring you into the negative, but having gold reserves will stop you from going bankrupt or having to take loans to stay afloat while the building is being made.

VICTORIA 3 STARTER TIPS: EXPLOIT YOUR NATURAL RESOURCES

All countries have natural resources and benefits that are exclusive to its states, so every time you start a game, take a moment to check out all of the state traits. These vary massively by country, and they give you bonuses for your buildings and goods produced in them. For example, several states in Sweden have the Scandinavian Forests trait, which increases the logging industry throughput, meaning you should focus these industries in those states to earn the benefits and maximise your economic output.

VICTORIA 3 STARTER TIPS: UNPRODUCTIVE TRADE ROUTES CAN STILL BE USEFUL

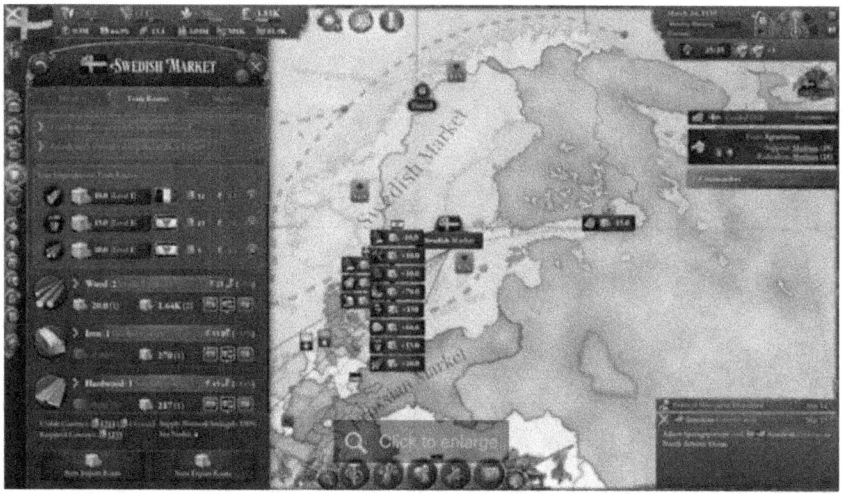

A sad reality of Victoria 3 is that you won't be able to produce all the goods you need in your own country, so sometimes you will need to import goods from other countries. This will bring more of that good into your country for pops to buy or buildings to consume during production.

Sometimes these trade routes can become unproductive and lose you money due to the tariffs involved, however, this doesn't always mean you should cancel them. If you can take the monetary hit and there are no other good trade routes to use, keeping the trade route to ensure the good is available in your country can be worth it, especially when it's vital in the production of other important goods.

VICTORIA 3 STARTER TIPS: UNIFY NATIONS TO EXPAND

The nations available during 1836 aren't the only nations that can exist in your game, and during the 100-year period that Victoria 3 takes place in, other nations may rise up due to unification. If you click the 'Cultures' button and select the 'Nation Formation' tab, it will list any potential nations you can form based on shared cultural heritage. This will allow you to expand your land, and consequently, gives you more states to build in and increase your GDP. Beware, you will take a Bureaucracy hit due to the new states you will be running, but you'll make up for this in the long if you manage your new nation correctly.

VICTORIA 3 STARTER TIPS: DON'T RESEARCH TECHS YOU ARE GETTING TECH SPREAD FOR

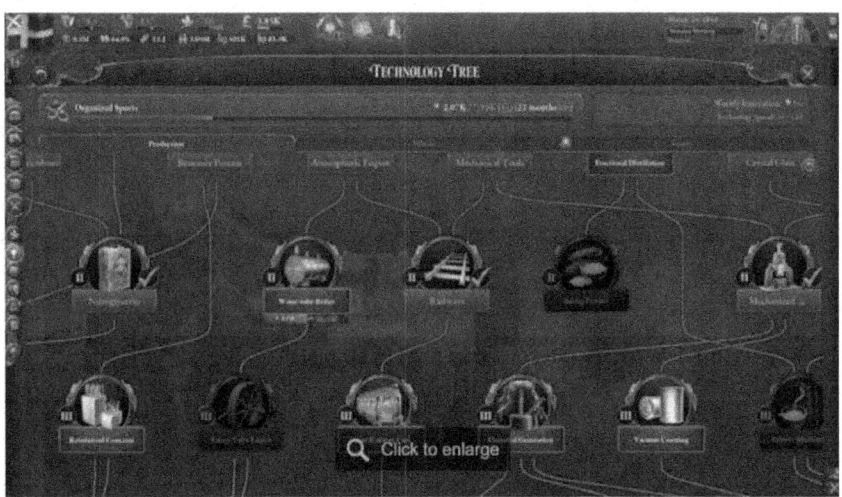

An important part of progression in Victoria 3 is the technology tree, which allows you to research new technologies in the Production, Military, and Society tabs. This will allow your nation to advance forward with more efficient production methods, military units, and laws, but you'll want to manage how you discover new tech. While you can choose a technology to research, you will also get technology spread from other nations that have technologies unlocked. So, if you're close to unlocking a technology due to technology spread, it may be worth researching another technology and getting them both at similar times.

Make sure you also know how to increase Bureaucracy in Victoria 3, which is the Capacity used for running the day-to-day aspects of your nation.

VICTORIA 3 STARTER TIPS: FOCUS ON BUILDING BASIC GOODS FIRST

The goods you build in Victoria 3 are split into four categories: staple goods, industrial goods, luxury goods, and military goods. While luxury goods can get you a lot of money - especially if other countries set up trade routes for access to them - you should focus on making sure your population's needs are met with staple goods to begin with. Goods like wood, basic clothes, grain, and much more are vital for running your country, and you won't get very far if these aren't plentiful for your population. After these basic needs are met, you can begin to specialise more with luxury goods and turn into a powerhouse producer of wine.

VICTORIA 3 STARTER TIPS: USE CONSUMPTION TAXES ON LUXURY GOODS

Consumption taxes are taxes collected on goods purchased by a population, increasing your revenue, while also raising the market price of the good. These cost Authority to establish, but if you have tonnes to spend, it may be worth enacting it on some of the more luxurious goods. Apart from the obvious monetary increase this can provide to your economy, the increase to the good's market price can also improve your GDP marginally, which comes with a tonne of useful effects.

That's all for our Victoria 3 starter tips, and now you should know some basic ways to benefit your nation and improve your Victoria experience.

BEST COUNTRIES TO PLAY IN VICTORIA 3

Our list of the best countries to play in Victoria 3 aims to cover all the bases, with a variety of nations that range from easy to hard. Victoria 3 lets you play any nation that existed during 1836 and take them all the way through the Industrial Revolution to 1936, while controlling aspects of their economy, government, diplomacy, population, and more. You're given the freedom to lead your nation in your own way, with gameplay open to a tonne of ideologies that can be put into practice. So if you're wondering what some of the best countries to play in Victoria 3 are, we've got you covered.

BEST COUNTRIES TO PLAY IN VICTORIA 3: GREAT BRITAIN

Great Britain is a classic choice for Victoria, as during the game's start date of 1836 the Pax Britannica was in full swing and Britain was the most powerful country on the globe. In many ways, it's one of the easier countries to play as in the game due to the huge GDP you begin with and the various overseas colonies, however, it's an enticing challenge to hold onto that power and stay number one for the whole 100-year playing period. There's also a fun achievement for converting Britain from a monarchy into an anarchist state that's worth attempting if you go for this starting nation.

BEST COUNTRIES TO PLAY IN VICTORIA 3: INDIAN TERRITORY

The Indian Territory is the only native nation you can control in Victoria 3, and during a game, the rest very quickly disappear as Mexico and the United States of America expand from each side of the North American continent. While in real life, the end of the American Civil War saw the Indian Territory brutally assimilated into the USA, you're given an opportunity to turn the tide and return North America to its natives. It's not something we suggest for your first playthrough as you'll need to master every mechanic available to you to fend off American imperialism, but this makes it the ultimate challenge for any player with big goals.

BEST COUNTRIES TO PLAY IN VICTORIA 3: EGYPT

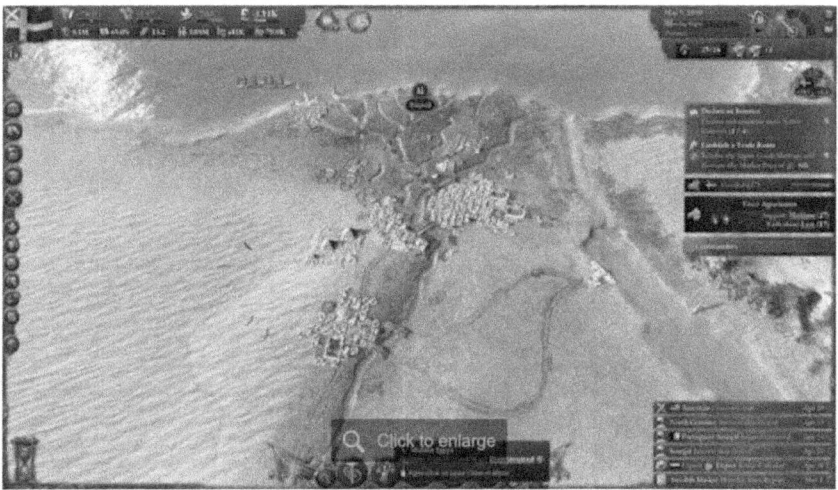

Egypt has persevered through everything thrown at them for over 5000 years, and you can help them personally thrive through the Industrial Revolution too. Egypt is relatively powerful during Victoria 3, but they are uniquely positioned as one of the few nations that can form Arabia and create a strong Muslim empire across North Africa and the Middle East. This is a big challenge to undertake and is considered one of the harder achievements in the game, but map painting is always a fun prospect in Paradox titles.

BEST COUNTRIES TO PLAY IN VICTORIA 3: CAPE COLONY

The Cape Colony is located in modern-day South Africa, and during the start of a Victoria 3 game, it remains a crown colony with very little autonomy over its own affairs. While it can be fun to start as a powerful nation and build even more power, it's arguably more rewarding to turn a nation from the conquered to a conqueror. Will you be able to successfully kickstart the revolution to remove Great Britain's control and claim your own destiny?

BEST COUNTRIES TO PLAY IN VICTORIA 3: GREAT QING

China as we know it didn't exist in 1836, but its land was ruled by the imperial dynasty of Qing until 1912 under an absolute monarchy. In-game, Great Qing is an unrecognised power, but with the highest GDP, several vassal states, and a third of the world's population, it's in a good place to be taken into the next era and earn its recognition as a great power. Playing as Great Qing is also a good tutorial for the diplomacy system in Victoria 3, as you need to challenge recognised powers to up your rank and status in the world.

That's all for our primer on the best countries to play in Victoria 3, and while there are many more nations worth your time, these were our personal favourites.

THE WHOLE WORLD AT YOUR FINGERTIPS

Victoria 3 is a grand strategy game where you're given the reigns of a nation from the starting year of 1836, and tasked with taking them all the way through the Industrial Revolution and the start of the 20th Century until 1936. Much of the game takes place during the age of imperialism, when the Pax Britannica was in full swing, and undeveloped nations across the planet were seen as ripe for the taking by European powers.

Much of the fun in the game is of your own making. There are no real objectives or end goals to speak of, and it's up to you to decide your own fate. Want to dethrone Great Britain as the world's foremost great power? Want to become the world's biggest producer of luxury clothing? Looking to just paint the map and form a modern nation? All of these and much more are perfectly valid options. It's a free form sandbox where you're only limited by your mastery of the game's systems and your imagination - plus there are specific achievements if you need a little push.

Aside from the differences highlighted above, Victoria 3 plays very differently from the other Paradox games, like Europa Universalis or Crusader Kings. Much of the game's mechanics revolve around the overall management of your nation, with a focus on economy, population, and politics. These systems facilitate building tall far more than any other grand strategy, and you can go the entire 100-year period without expanding outward, and still feel accomplished due to how much you can affect change within your own borders.

HOW TO BE AN OMNISCIENT RULER

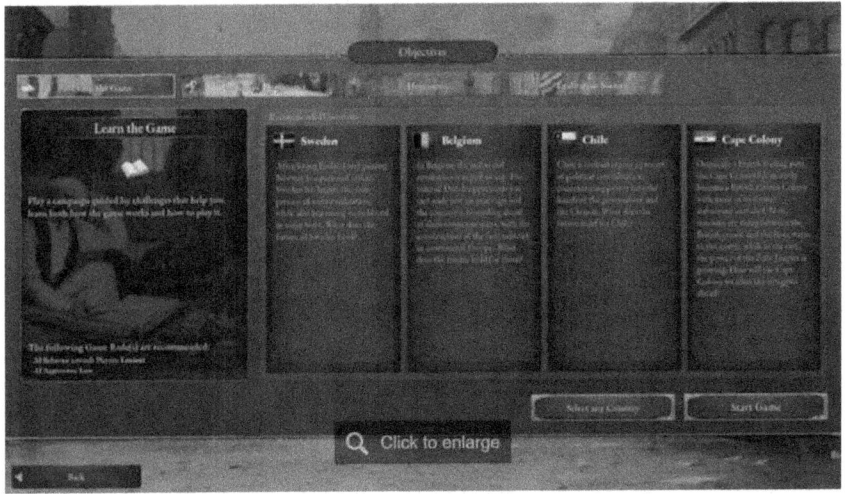

It's fair to call Victoria 3 a very complex game, but it's not incomprehensible. I had no experience playing the previous Victoria games before this, but Paradox's robust tutorial provided a great introduction to the game's base systems, which shows how far they've come at actually teaching people how to play their games.

When you select to start a new game, you're given the option to choose to learn how to play, and then presented with four countries: Sweden, Belgium, Chile, and Cape Colony. These are guided campaigns that teach you the basics, and periodically provide you with objectives to learn about the game's more intricate systems. These specialised campaigns essentially let you learn on the job, and encourage you to experiment with mechanics until you have a better understanding, without the risk of everything blowing up in your face by nerfing the AI. But if you want to take off the training wheels, you'll need to start another campaign, which is where the real challenge begins.

Many of the decisions you can make during a game of Victoria 3 are effected by four key resources: Bureaucracy, Authority, Influence, and Money. The first three of these are called capacities in-game, which are measures of your overall capabilities. These affect the day-to-day operations of your country, your ability to make change, and your ability to effectively take part in diplomacy, respectively. Money on the other hand is used for a myriad of purposes, like paying government and army wages, and paying for construction materials while building buildings.

These resources also feed into other aspects of the game. For example, you will need Bureaucracy if you want to increase the power of institutions like education, which will have a knock on effects toward increasing literacy. Learning what each resource effects is key to mastering how to make the most out of each action you take, and part of the satisfaction of learning Victoria's systems is seeing how everything begins to connect together.

LINE GO UP SIMULATOR

As I said earlier, economic management is a principal feature of Victoria 3, and perhaps the thing you will spend the most time on. Each country has their own simulated economy, which is based on the supply and demand of individual goods that can be produced in buildings. If you want more tools for example, you will need to build workshops or export that good into your nation. Each building requires input goods which are used to produce output goods, and managing each building becomes a balancing act of making sure they're still productive to your economy. All buildings can be expanded to increase the workforce and overall output, but you can also change production methods based on which technology you research, or which system of economic management you prefer.

This is where the Victoria 3's strategic management gameplay really begins to shine, and once again, where you can see how the various mechanics play into one another. If you want to change the production method for your wheat farm, you'll need to make sure you have enough of the required input goods, or enough educated people in your population to take on the different jobs that are required. It's micromanaging at its finest, and I spent hours poring over the different screens that represent the market and the buildings to see what changes I could make to improve productivity by even the smallest margins.

The next principal area of management revolves around your population and the government. Your nation's population is organised into groups called Pops, which are swathes of people who share professions, and people who share the same culture and religion. These Pops are also organised into interest groups, which are political factions who share ideologies and desires for how the nation is governed. These are key for a few reasons, and they also present an obstacle for any changes you may want to make in your country.

Interest groups relate to the laws you can pass to set the rules for how your nation operates. If you want to push your nation toward a less autocratic model, you will need to perform some political manoeuvring by getting an interest group that supports voting and elections into power, and then pass a law that enables this system. This is a central conflict throughout any game of Victoria 3, and another huge part of the strategic puzzle. Figuring out how to get interest groups into power that represented the middle and lower class more, was a eureka moment for me during my playthrough, and enabled me to pass laws that opened up more gameplay avenues.

THE GRANDEST OF STRATEGY

Victoria 3 is a premiere example of the unique gameplay opportunities presented by grand strategy games, making for a playground where the swings are replaced by GDP graphs, and the slides are swapped out with law reform mechanics. It's a niche game that at its core is about enacting social and political change through close internal management of your nation, and nothing on the market comes close to replicating it.

HOW TO INCREASE GDP IN VICTORIA 3

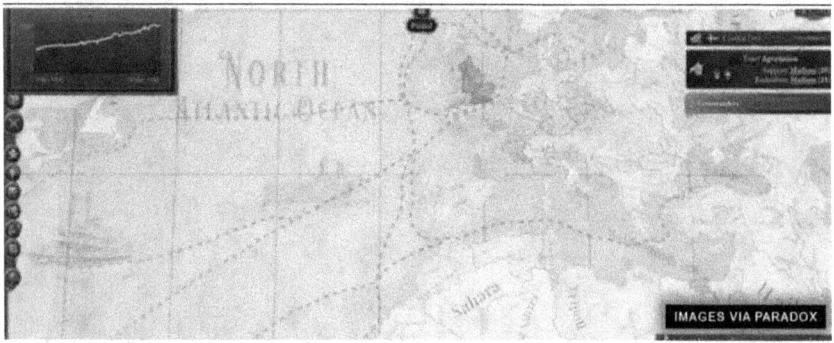

Learning how to increase GDP in Victoria 3 is a vital aspect of growing your nation, as it directly affects your prestige and overall standing amongst the other nations of the world, along with your potential for minting new currency. Controlling and managing your economy is one of the most important parts of playing Victoria 3, and many of the decisions you make will directly contribute toward your overall Gross Domestic Product (GDP). So if you want to know how to increase your GDP in Victoria 3, and gain all the benefits that come with it, we've got you covered.

HOW TO INCREASE GDP IN VICTORIA 3

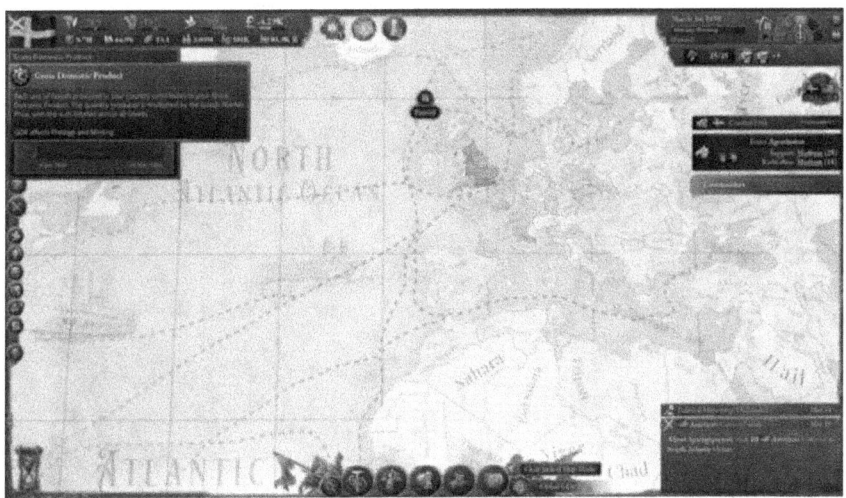

GDP is a monetary measure of the market value of all your produced goods and their market prices, which can be compared against other nations to see who has the more powerful economy. It's one of the most important indicators of your nation's success in Victoria 3, and it directly affects your prestige, with a larger GDP earning you more prestige. It also affects minting - which is a form of income generated through creating new money - as the larger your GDP, the more new money that can be minted without compromising the economy.

The equation for calculating your GDP in Victoria 3 is fairly simple. Essentially, the game takes the total amount of a good produced, and multiplies it by the market price of that good. It performs this calculation for all the goods produced in your nation, and then totals them together to work out your GDP.

The key to increasing your GDP in Victoria 3 is producing more goods. The idea behind this is simple, but in practice, it can be difficult and is dependent on multiple factors. Firstly, you should take a look at your nation's natural resources and see which goods have the best throughput. Throughput is a measure of how effectively your workforce can turn input goods into output goods.

You can see your nation's natural resources by clicking a state in your nation, and scrolling down in the Overview window to see its state traits. For instance, in the above screenshot, the state of Gotaland has the Scandinavian Forests trait, which provides a +20% modifier to logging industry throughput. This means any logging building that's built in this state will be more effective at creating wood.

Starting with the basics of your own goods economy can also be massive for your GDP. Making sure your population is well-fed with goods like grain or meat, making they sure have basic clothing, and ensuring you have wood and tools to continue to build more buildings, is essential throughout the game. If you don't produce enough of these goods, your people and your nation will suffer, so even if you intend to become the biggest producer of silk down the line, you won't be able to get there without the basics.

Another good way to move toward increasing your GDP, is by exploiting gaps in the goods market. By clicking the Market button on the left of the screen, you can see a full breakdown of all the goods produced in your market, including the sell orders, buy orders, the total balance of the good, and the market price. Clicking another nation when this menu is up will also let you take a look at their market and all the finer details. If you can find a good in their market that's got a low balance, producing more of it in your market can cause their AI to establish import trade routes with you, or you could export the good yourself. Along with making you money through tariffs, this will contribute toward your GDP.

You can also exploit this demand by being the first to produce a specific good. If you have a technological advantage and are the first to research Atmospheric Engines, for example, you can set up a motor industry and become the primary producer of engines. Once other nations start to create a demand for engines, you will be there to fill that niche, as you have a head start in production of the necessary good.

There's a lot of micromanaging you can do when growing your economy in Victoria 3, so experimenting with your production industries can be an important way of learning how to increase your GDP. There's no one way of increasing GDP, but with a combination of the tips above, and learning about the economy you inherit when starting a game, you should be raising your prestige in no time.

That's all for our breakdown of how to increase GDP in Victoria 3, and now you should have an idea of the best ways to make your GDP line go up.

HOW TO INCREASE LEGITIMACY IN VICTORIA 3

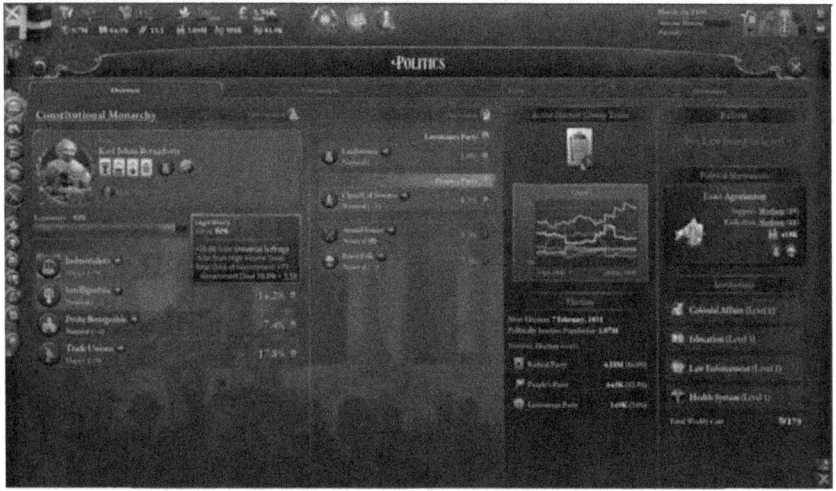

If you want a successful government, you will need to know how to increase Legitimacy in Victoria 3, as it directly relates to passing legislation and low Legitimacy can make your job of stewarding your nation into a brighter future even harder. Victoria 3 is a challenging grand strategy that lets you control a country through the turbulent Industrial Revolution, forcing you to think about your economy, diplomacy, population, and more. So if you're looking for an explainer of how to increase Legitimacy in Victoria 3, we've got you covered.

HOW TO INCREASE LEGITIMACY IN VICTORIA 3

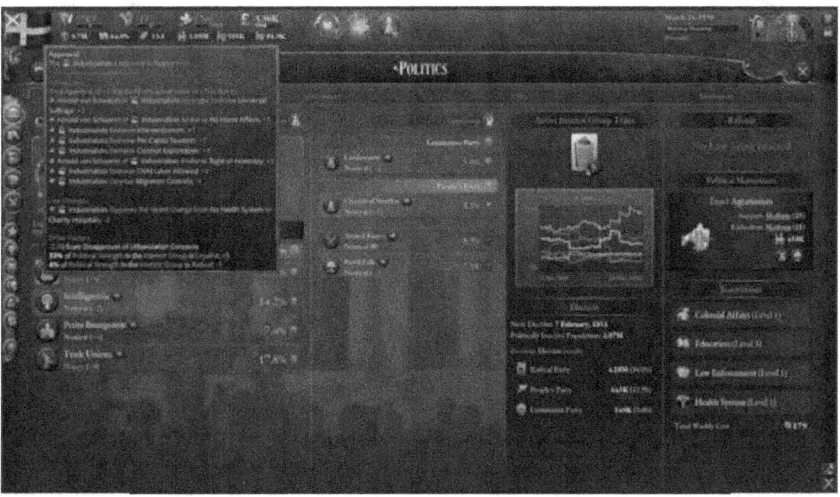

In Victoria 3, your Legitimacy is a measure of how much the interest group's in government fit with the country's established laws, meaning if you have laws that are supported by your government, your Legitimacy will be higher, and vice versa. There are other aspects that affect your Legitimacy, such as raising or lowering taxes, and passing specific laws can give you different Legitimacy modifiers regardless of who's in power.

You'll want your Legitimacy to be as high as possible, as low Legitimacy will increase the time it takes between attempts to enact new laws. Generally, this will make it harder for your government to actually get anything done, causing you to lag behind other nations as they modernise quicker.

A good way to get a quick boost of Legitimacy is to lower taxes, but this can lead to issues regarding your nation's finances in the long run, so this method should only be used when you can take the financial hit.

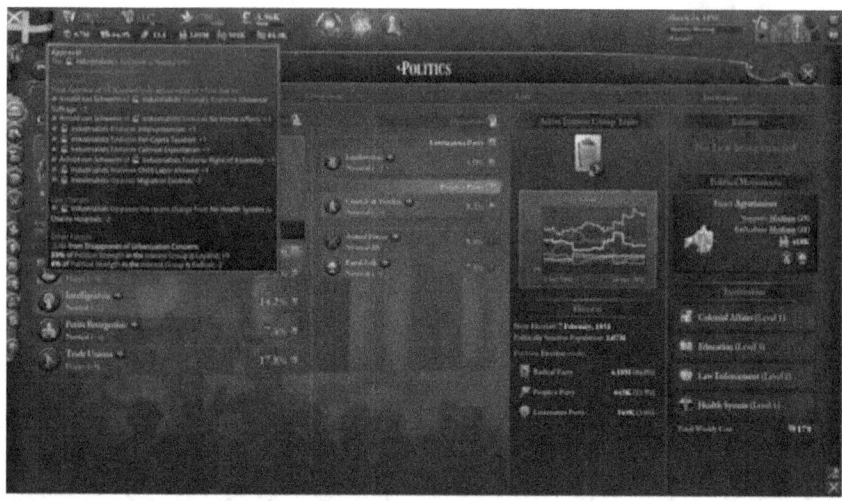

A better way of raising your Legitimacy is to take a look at the specific ideologies of the interest groups in power, check what laws they support, and attempt to enact them. You can do this by selecting the interest group from the Politics Overview screen. Hovering over their happiness modifier will also directly tell you which laws they currently approve or disapprove of.

Below, we'll list all of the laws that can affect Legitimacy, along with their specific modifiers:

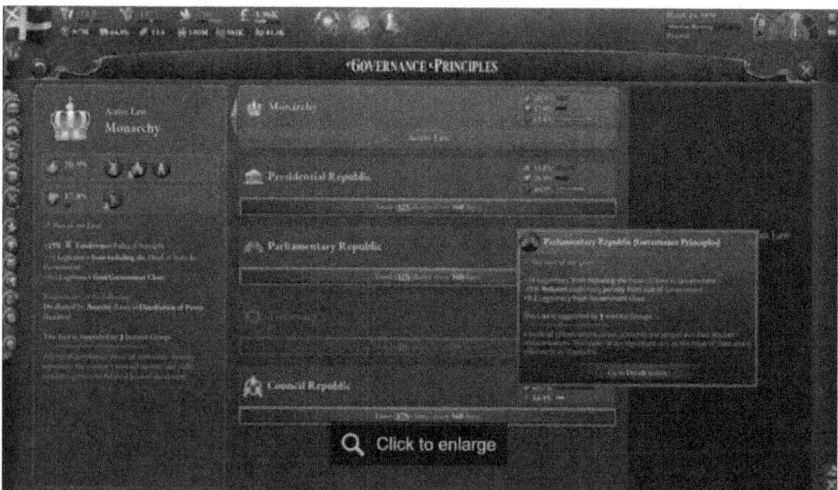

Governance Principles

Monarchy

- +20 Legitimacy from including Head of State in government
- +0.1 Legitimacy from government clout

Presidential Republic

- +15 Legitimacy from including Head of State in government
- -10% reduced Legitimacy penalty from size of government
- +0.2 Legitimacy from government clout

Parliamentary Republic

- +10 Legitimacy from including Head of State in government
- -25% reduced Legitimacy penalty from size of government
- +0.2 Legitimacy from government clout

Theocracy

- +20 Legitimacy from including Head of State in government
- +0.1 Legitimacy from government clout

Council Republic

- +0.4 Legitimacy from government clout

Distribution of Power

Autocracy

- +30 Legitimacy from including Head of State in government.

Oligarchy

- +25 Legitimacy from including Head of State in government

Landed Voting

- +20 Legitimacy

Wealth Voting

- +20 Legitimacy

Census Suffrage

- +20 Legitimacy

Universal Suffrage

- +20 Legitimacy

Anarchy

- +10 Legitimacy

That's all for our breakdown of how to increase Legitimacy in Victoria 3, and now you should have an idea of how to keep your Legitimacy high and enact laws quicker.

HOW TO INCREASE INFLUENCE IN VICTORIA 3

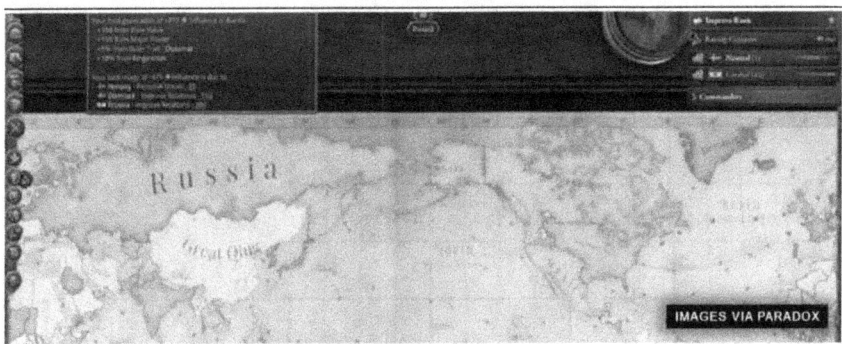

You will need to know how to increase Influence in Victoria 3, as it directly affects your relationships with other countries and allows you to make plays on the world stage for the benefit of your nation. Victoria 3 is the latest grand strategy from Paradox, throwing players into the turbulent Industrial Revolution and tasking them with governing a nation through politics, economy, diplomacy, and more. With such a connected world of trade, your Influence over nations will be vital, so check out how to increase Influence in Victoria 3.

HOW TO INCREASE INFLUENCE IN VICTORIA 3

Influence is one of the three major Capacities in Victoria 3, which are measures of your country's overall capabilities. The Influence Capacity is mainly used for things outside the realm of your nation, such as making Diplomatic Actions or establishing pacts with other countries.

Like every other Capacity, you begin with a base value of Influence, but you gain extra output based on your rank. Every country in the game has a rank, which is determined by comparing their Prestige value to all other countries. Prestige is a measure of your overall country, based on GDP, military power projection, and country tier, though you can earn bonuses for it by exploring Arts technologies and being a world leader at producing a market good.

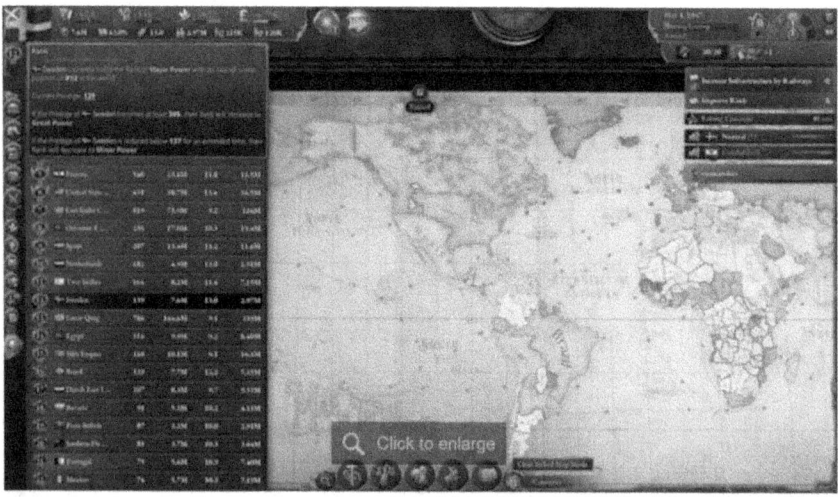

Each rank has a flat Influence increase associated with it, and the higher your rank, the more Influence you gain. Below are the different ranks available, along with the Influence you gain for reaching it:

- Great Power: +1000 Influence

- Major Power: +750 Influence
- Minor Power: +600 Influence
- Insignificant Power: +500 Influence

There are also Unrecognised versions of each rank, which means they are perceived as inferior to their actual rank, and restricted in what ranks they can achieve, though it doesn't seem to affect Influence gain.

Aside from the Influence you earn based on your rank, there are some other ways you can earn smaller Influence modifiers. Some ruler traits do this, such as the Diplomat trait which gives +5% Influence. There are also interest group traits which affect your Influence. If the Landowners interest group has +5 happiness, they will activate the Family Ties trait which gives you +10% Influence, while if the Petite Bourgeoisie have -5 happiness, they will activate the Xenophobia trait which gives you -10% Influence.

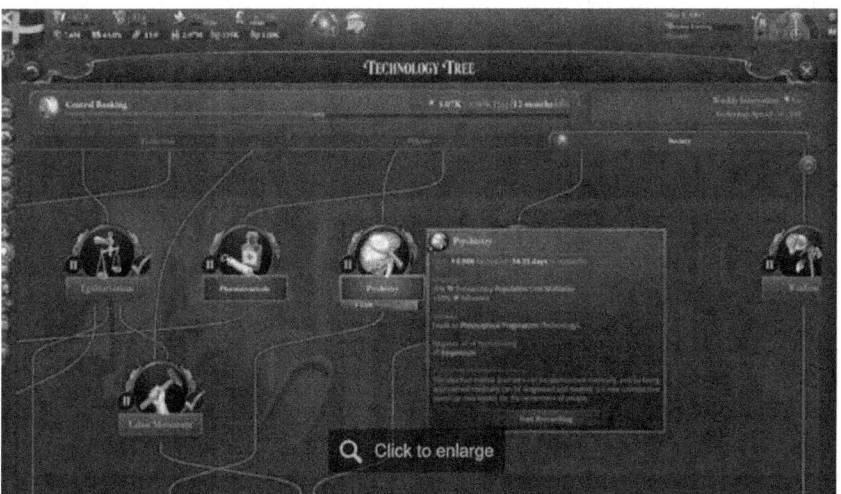

Finally, there are several technologies on the Society Technology Tree which give you modifiers to Influence. Below, you can find the name of the technology, along with the Influence modifier:

- Empiricism: +10% Influence
- Psychiatry: +10% Influence
- Philosophical Pragmatism: +10% Influence
- Psychoanalysis: +10% Influence
- Behaviorism: +10% Influence

We also thought it would make sense to list the technologies that provide Prestige bonuses, as it directly affects your rank, and thus your Influence. These can also be found in the Society Technology Tree, and are:

Romanticism: +5% Prestige

Realism: +5% Prestige

Organised Sports: +10% Prestige

Camera: +5% Prestige

Film: +5% Prestige

That's all for our primer on how to increase Influence in Victoria 3, and now you should have an idea of how to boost it, so you can make diplomatic choices on the world stage.

HOW TO INCREASE AUTHORITY IN VICTORIA 3

You will need to know how to increase Authority in Victoria 3 if you want to make ruling your nation easier, and enjoy the benefits of a population that won't rise up against you. Victoria 3 challenges players to make a name for their nation during the tumultuous Industrial Revolution, giving you 100 years to sway the politics, economy, and population of your country. If you want to become a world power, then you will need to have plenty of Authority, so check out how to increase Authority in Victoria 3.

HOW TO INCREASE AUTHORITY IN VICTORIA 3

Authority is one of the three major Capacities in Victoria 3, which are measures of your country's overall capabilities. The Authority Capacity is used to make internal changes to your country, such as issuing decrees, managing interest groups, and collecting consumption taxes on goods.

You'll always begin with a base value of Authority, but the main way to increase it is through enacting specific laws, which reward you with a flat Authority increase or percentage increase. Generally, laws that increase the liberties of your population, such as wider distributions of voting power, will lower your Authority output, so it's worth keeping a good balance until you can get lots of Authority from other sources. Below are all the laws which increase or decrease Authority, and their output, broken down by law group.

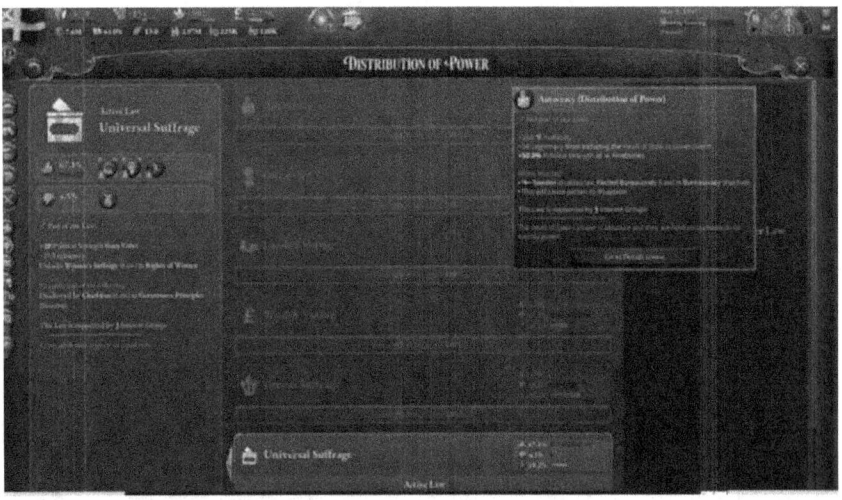

Distribution of Power

- Autocracy: +250 Authority
- Oligarchy: +200 Authority
- Landed Voting: +150 Authority

- Wealth Voting: +100 Authority
- Census Suffrage: +50 Authority
- Anarchy: -50% Authority

Citizenship

- Ethnostate: +200 Authority
- National Supremacy: +200 Authority
- Racial Segregation: +100 Authority
- Cultural Exclusion: +50 Authority

Church and State

- State Religion: +200 Authority
- Freedom of Conscience: +100 Authority
- Economic System
- Command Economy: +25% Authority

Trade Policy

- Isolationism: +50% Authority

Free Speech

- Outlawed Dissent: +200 Authority
- Censorship: +100 Authority
- Right of Assembly: +50 Authority

Aside from increasing your Authority with laws, you can also earn it from a few other sources. Having a popular ruler will increase Authority by a flat amount, and they can also have traits, such as

Political Operator, which gives your Authority a +5% boost. The Church interest group also has a group trait that's activated when they have five happiness, which gives you +10% Authority.

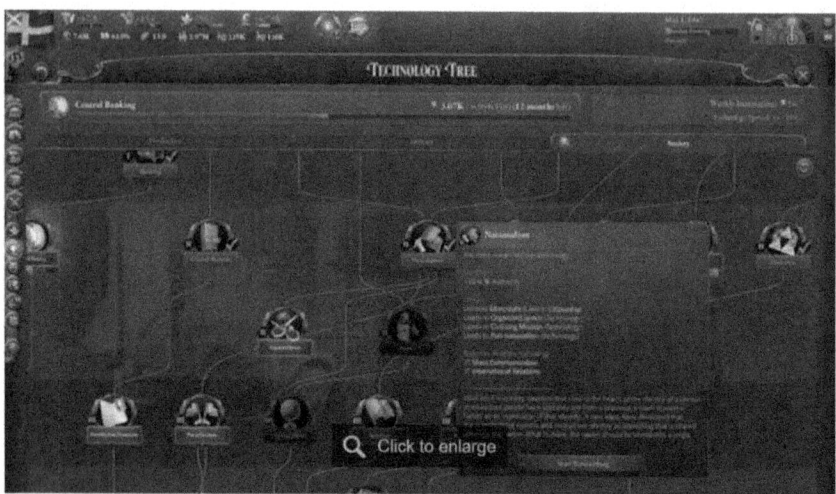

Lastly, some technologies from the Society Technology Tree will also provide Authority multipliers which are worth seeking out, especially when you want to enact more liberal laws for your nation. Below are the Technologies that change your Authority, along with their modifier:

- Mass Communication: +10% Authority
- Nationalism: +10% Authority
- Pan-nationalism: +10% Authority
- Political Agitation: +10% Authority
- Mass Propaganda: +10% Authority

That's all for our breakdown of how to increase Authority in Victoria 3, and now you know how to gain more Authority through laws, along with modifiers from traits and technology.

HOW TO INCREASE BUREAUCRACY IN VICTORIA 3

You will need to know how to increase Bureaucracy in Victoria 3, as it's one of your most important resources for progressing your government and nation forward. Victoria 3 has a tonne of interconnecting systems, and once you first get going it can be quite overwhelming to learn. However, once you've wrapped your head around the basics of running a nation, you'll be a major power in no time. So check out our breakdown for how to increase Bureaucracy in Victoria 3.

HOW TO INCREASE BUREAUCRACY IN VICTORIA 3

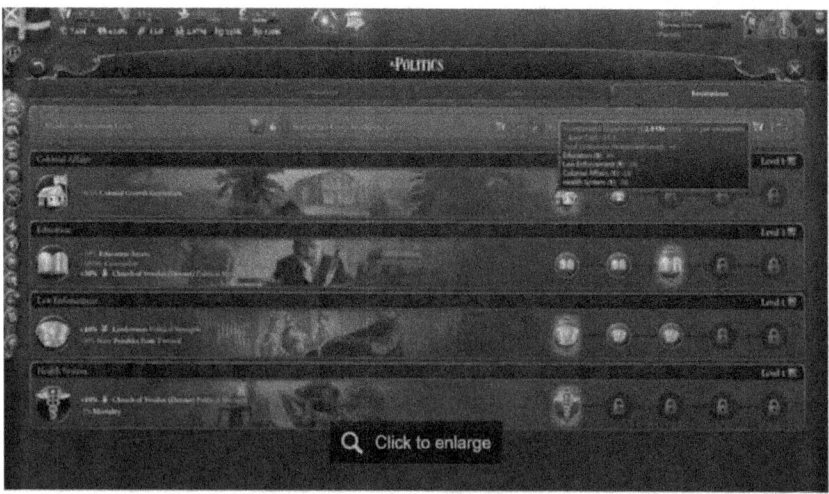

Bureaucracy is one of the three major Capacities in Victoria 3, which are measures of your country's overall capabilities. The Bureaucracy Capacity is specifically related to maintaining your nation's day-to-day operations and administrative cohesion, and is used for the upkeep of your population, states, Institutions, commanders, and trade routes.

While you begin with a base value of Bureaucracy, the main way of increasing it is through building and upgrading Government Administration buildings. These are urban building types which provide +50 Bureaucracy for each level, along with increased taxation capacity, but they require paper to maintain, so you'll want a steady balance of paper if you're looking to expand your admin buildings.

If you have a surplus of Bureaucracy, it will increase the efficiency of constructing new buildings. If it slips into the negative, you will get a large penalty, so it's imperative to always make sure you have a positive output of Bureaucracy. This also ensures you can spend it during emergencies, such as establishing a trade route for an important good that you have a shortage of.

The only other method of improving your Bureaucracy is through the use of a specific trait. If the Petite Bourgeoisie interest group has a happiness of at least five, they will activate the Middle Managers trait, which gives a flat 10% boost to all your Bureaucracy output. This is a powerful trait to have activated, and if you're struggling to maintain a large amount of Bureaucracy, it may be worth appeasing their wants.

That's all for our primer on how to increase Bureaucracy in Victoria 3, and now you should have an idea of how to maintain a positive Bureaucracy balance.

HOW TO INCREASE LITERACY IN VICTORIA 3

Knowing how to increase Literacy in Victoria 3 is imperative to having an educated population who can read and write, but it's not immediately clear how to increase it. Victoria 3 lets you manage a nation during the Industrial Revolution, challenging you to adapt its government, economy, and people to the times, while staying competitive on the world stage. Your Literacy rate plays a huge role in pushing your nation forward, so take a look at how to increase Literacy in Victoria 3.

HOW TO INCREASE LITERACY IN VICTORIA 3

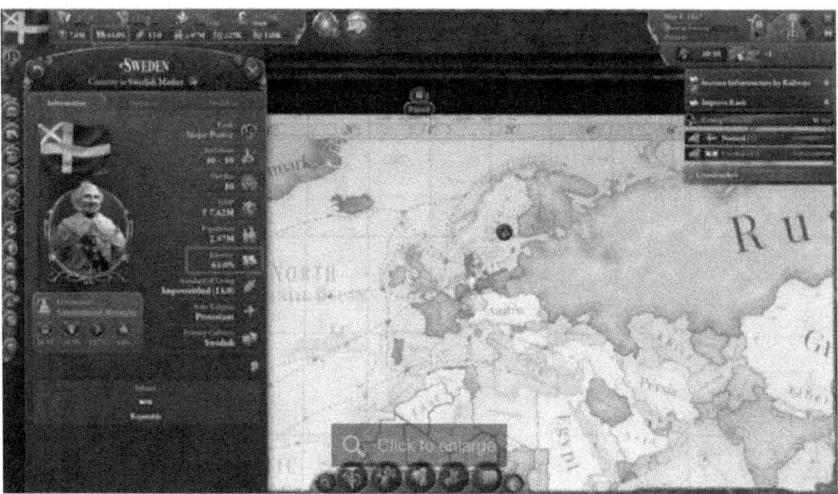

In Victoria 3, your Literacy dictates what percentage of your total population can read and write, which is important for a few different reasons. Firstly, if your population isn't literate, they won't be able to take part in certain professions that require qualifications, which will stunt your workforce and stop you from developing more advanced goods over time. Literacy also directly affects your Innovation, which is a value that determines how long it takes to research technologies and receive technology spread. Essentially, if you have a low Literacy rate, your nation will suffer when it comes to progressing and developing further.

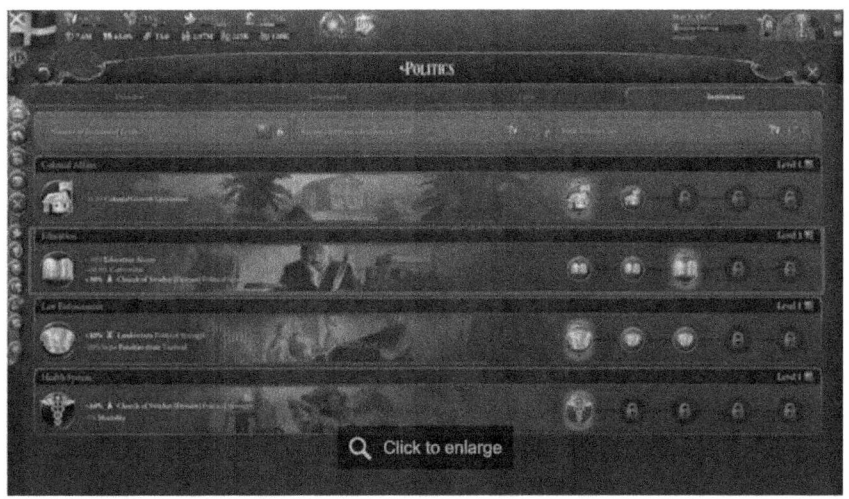

The main thing that affects your Literacy rate is your nation's Education level, and increasing this will cause your population to become more literate over time. Education is an Institution, and each level you put into it will increase Education access for your population. To access this, select the 'Politics' icon on the left menu, and then click the 'Institutions' tab. On this screen, (pictured above) you will be able to see your Education level, and the direct effect that it has on your population.

To gain the Education Institution so you can start putting points into it, you will first need to enact an Education System law. The first law, No School, will disable the Education Institution, but selecting Religious Schools, Private Schools, or Public Schools will enable it. Public Schools also increases Education access, which is vital for boosting Literacy.

You will need to max out your Education Institution to level five if you want to reach 100% Literacy for your population, but you will have to take some specific steps to reach the fifth level. It's also worth noting that each level of Education costs Bureaucracy, so you will need a lot to max out Education. Firstly, you will need to unlock the following technologies from the Society Technology Tree:

- Rationalism
- Empiricism
- Dialectics
- Human Rights

The first three of these technologies allow you to put one level into the Education Institution and will get the ball rolling for your Literacy rate. Human Rights don't directly affect your Education, but it will allow you to change the laws for Children's Rights, which is the next step for boosting Literacy.

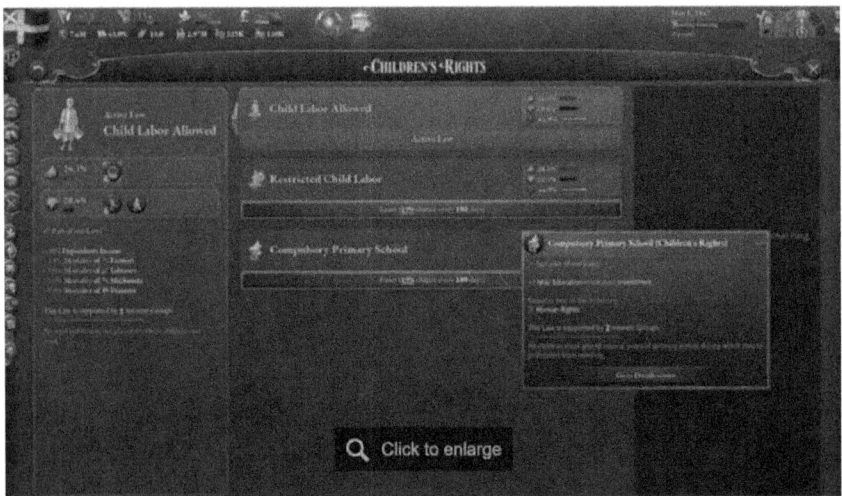

Once you have three levels in the Education Institution, you will need to pass a specific law to gain access to the final two levels and max out Education. We have a guide on how to pass laws in Victoria 3 if you're not sure of this next step. Navigate to the 'Laws' tab on the Politics screen, and select the Children's Rights law group. Passing Restricted Child Labour will let you put one point into Education, while Compulsory Primary School will allow you to max out Education.

Once the Education Institution is maxed out, your Literacy rate won't just jump up to 100% straight away, but as all children will now be literate, over time your Literacy rate will begin to climb.

That's all for our explainer of how to increase Literacy in Victoria 3, and now you know how to create a fully literate country through the use of Institutions and laws.

VICTORIA 3 TIMESPAN: WHAT IS IT?

You may be wondering about the Victoria 3 timespan, and how long a full game of Victoria takes to complete. Victoria 3 is the latest grand strategy from Paradox, which is known for deep and complex titles like Europa Universalis and Crusader Kings. The Victoria series differs from these quite a bit in terms of gameplay, and with the more modern time period it takes place in. So if you need to know what the Victoria 3 timespan is, we've got you covered with our explainer below.

VICTORIA 3 TIMESPAN

The Victoria 3 timespan is exactly 100 hundred years, running from 1836 to 1936, which takes you through the Industrial Revolution up until shortly before the outbreak of WWII. This is the second shortest timespan of the Paradox grand strategy franchises, and is exactly the same as the previous Victoria titles, so fans should be accustomed to the length of time.

A whole century is more than enough time to reach the goals you may be aiming for when starting a game of Victoria 3, and you'll have the chance to replicate the rapid modernisation this period saw in real life. While it's possible the timespan could be extended in the future with expansions or updates, it's likely that this will be set in stone for the game's life cycle. However, there will certainly be mods which change things up at some point if you are desperate to play for longer.

That's all for our explainer of the Victoria 3 timespan, and now you know how many years a Victoria 3 game runs for.

VICTORIA 3: HOW TO MANAGE YOUR CONSTRUCTION

If Pops are your country's lifeblood in Victoria 3, buildings are the arteries they live and work in.

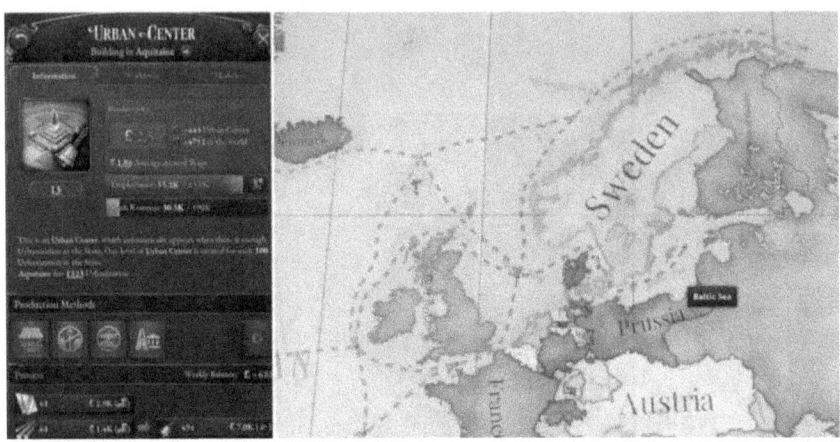

If Pops are the lifeblood of a nation in Victoria 3, then buildings are the arteries they pump through. Knowing how, when, and what to build is an important thing to get a handle of in most strategy games, and Victoria is no exception - managing the makeup of your states is a crucial skill to learn.

It is from buildings that you'll exploit your land and your states, how you move your population into gainful employment, and how to connect your peoples both to each other and with the outside world. It's also easy to muck things up - overexpansion is a curse in a game where money is everything, and plunging your Pops into debt as a result of poor planning is a surefire way to get a revolution on your hands.

Types Of Building

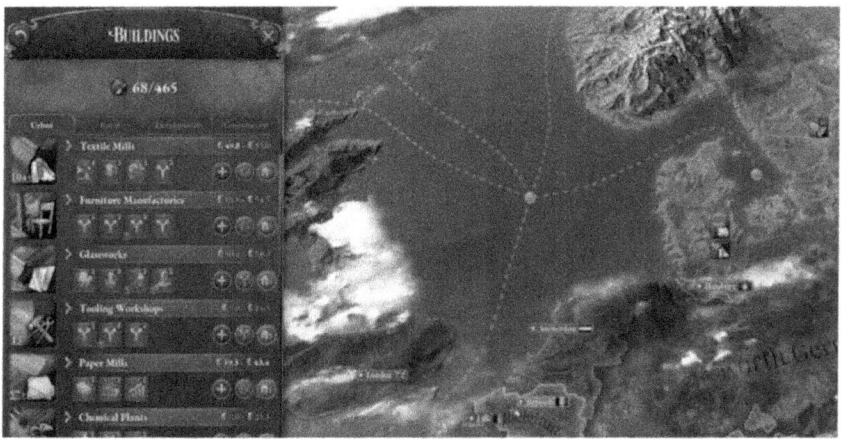

In Victoria 3, there are three different types of buildings. Rural buildings are used primarily to harvest resources from the land - the Rural buildings you can construct in any one state largely depend on the state's natural resources. You won't be able to build a Logging Camp in a state that has no forests, for example. In addition to these typical resource buildings, there's also a certain amount of Arable Land attributed to a state, and here you can build farms or leave them as Subsistence Farms - more on that later.

Conversely, Urban buildings are a little different. Like most other buildings, they'll take in Goods and turn them into different, usually more expensive Goods, but they also contribute a certain about of Urbanization to a state. When a state's Urbanization hits a new multiple of 100, it'll expand a building called an Urban Center, which cannot be expanded in any other way. Urban Centers are staffed like other buildings and represent markets and services, and they are key to maintaining a healthy economy. Similar to Urban Centers are Trade Centers, which grow as you focus more effort and money into trade routes and also employ lots of Pops.

Development buildings are a little more unique - these buildings are crucial to the running of your country but do little in the way of making money. These buildings include those used in war, those used for moving goods and Pops around, and those used for the construction of all other buildings.

Arable Land And Subsistence Farming

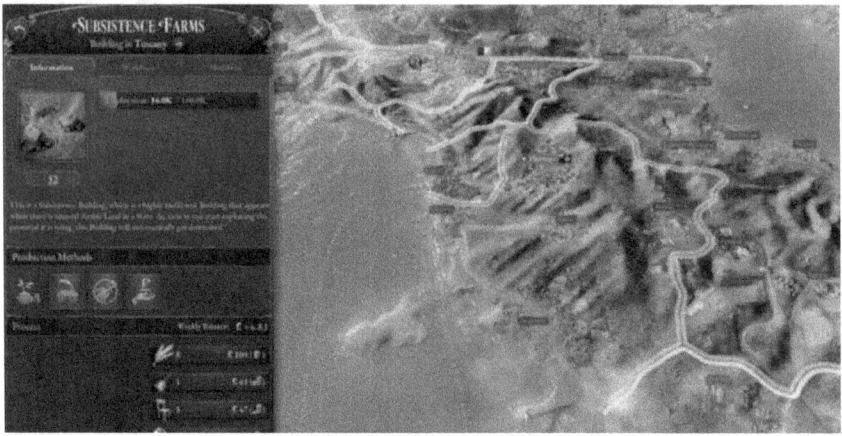

All states have a predetermined amount of Arable Land for use at the beginning of the game. When you build or expand a Farming building in a state, you use up one unit of that Arable Land, whether it's a Livestock Ranch, Rye Farm, or one of the many plantations available in the game. These all fall under the Agriculture heading, which is a subset of Rural.

Unused Arable Land is automatically used for Subsistence Farming by peasants. Subsistence Farms produce a variety of goods that the peasants will use to fulfill their meager needs, but they'll also sell some of it and let it enter the national market. The issue is that Subsistence Farms are bad at both providing for needs and for profit, so it's a good idea to draw as many peasants into gainful employment as possible. A good way to do this is to use up a lot of your Arable Land, but also to make sure there's enough entry-level employment opportunities available.

Urban Centers

Urban Centers pop up in your states after building a certain amount of Urban buildings. Specifically, a state's Urban Center will increase in size by one level for every 100 Urbanization it has, and Urbanization is a resource generated in most buildings, but mostly in Urban buildings.

You won't find Urban Centers as normal building tiles in the Buildings menu of a state - instead, there's a bar at the top that tells you its Urban Center's size. Click on this to go to the more traditional building screen that lets you know the details of its employment and also alter its production methods.

Urban Centers are important as they are the primary producer of Services, a Good that cannot be imported and is important for the needs of wealthier Pops.

If you want to increase the size of your Urban Centers, build and expand the buildings that contribute the most Urbanization: Government Administration, Motor Industries, Power Plants, Electrics Industries, and most other Urban buildings. Naturally, it's easier to raise your Urbanization as you get later into the game.

Trade Centers

Trade Centers, like Urban Centers, cannot be built directly. Instead, they pop up as a result of your trade routes. They'll be constructed automatically in the states responsible for such trades, and contribute a hefty amount of Urbanization to a state.

Trade Centers don't produce anything or consume anything like other buildings, but they will employ a large amount of Pops and are how you can contribute massively to the economy through trade - both through tariffs and through enriching your Pops.

Government Administration

Government Administration is a special Urban building that doesn't produce a typical resource like other buildings do. Instead, Government Administration buildings raise two things: Bureaucracy and Taxation Capacity.

- Bureaucracy is very important - you use it to promote generals, sustain your incorporated states, supply your government institutions, and support your trade routes. Without a positive Bureaucracy, your economy tanks in a big way and you'll be unable to keep your country running smoothly.

- Taxation Capacity is also very important. Taxes are where you'll get a very large proportion of your income, allowing you to fund construction projects and everything else required for running a nation. Every point of Taxation Capacity can sustain 10,000 Pops and their taxes - if your population grows larger than your Taxation Capacity, you're letting precious money get wasted into the ether.

The best way to increase both of these variables is through

Government Administration buildings, and since Taxation Capacity is a state-specific stat, you'll likely want them in most every state.

Infrastructure

There are two buildings in the game that provide a state with Infrastructure, which is crucial to maintaining a healthy economy and supporting your supply chains. These are Ports and Railways.

- Ports give your states Infrastructure but that's largely a fringe benefit. Ports are primarily used to increase your number of Convoys, which is required and important for overseas trade.

- Railways, on the other hand, are vastly better at providing states with Infrastructure. Note, however, that improving the production methods reduces the amount of Infrastructure they provide in exchange for a higher Transport production.

Construction Sectors

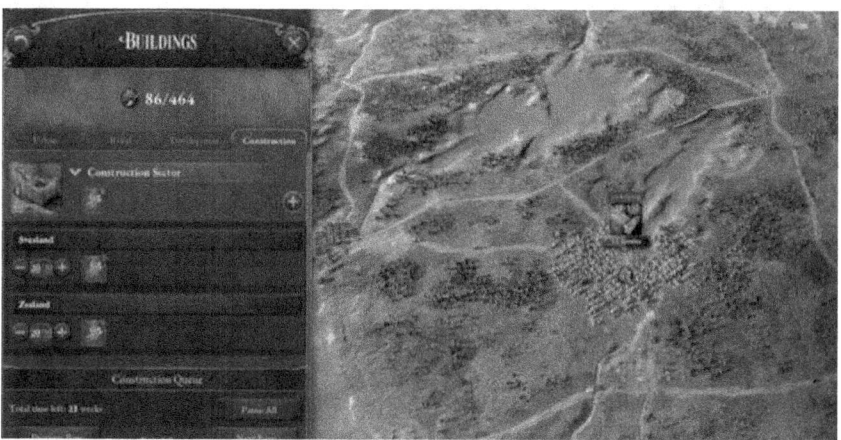

Construction Sectors are the final type of building we'll look at - they contribute Construction, which is a representation of your ability to, well, construct buildings. The higher your Construction, the quicker you'll be able to build and expand your buildings.

It is important to note that Construction is incredibly expensive. While it is possible and even recommended sometimes to queue up many months, even years, of construction in advance, construction requires lots of Goods and money. This can put a strain on your economy.

It is for this reason that we recommend being extra thoughtful when it comes to large building projects and being liberal with the Automatic Expansion feature, especially when you don't have much gold in reserve.

When To Expand Buildings

A good rule of thumb is to expand buildings when they reach the same prerequisites that they would for Automatic Expansion, to whit:

- The building's cash reserves being over 95 percent full.

- The state in question having sufficient market access thanks to their infrastructure.

In addition to these, try to only upgrade buildings that are fully or nearly fully employed. These are the buildings that are doing very well and are the ones that are most likely to be able to handle the larger employment maximums and Goods production.

There are some times that you would prefer to hold off expanding, though. If it uses a Good that you struggle to produce or produces a Good that you have a giant surplus of, it wouldn't be beneficial to expand just yet.

That said, there are some buildings that are almost always good candidates for expansion - these include buildings that harvest raw materials, such as Iron Mines and Whaling Stations, and buildings that use Arable Land.

VICTORIA 3: HOW TO RAISE AND LOWER THE PRICE OF GOODS

Knowing how to manipulate the price of goods is something you'll need to learn if you want to manage a successful country in Victoria 3.

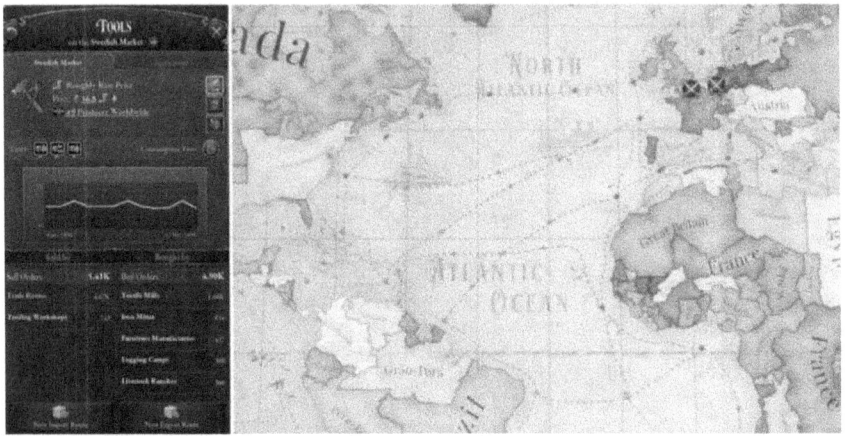

One of the most important mechanics in Victoria 3 is its incredibly deep economic simulation. The supply and demand of Goods is something you'll have to learn to manage from the word 'go' in this game and knowing how to manipulate prices is one of the keys to success.

It can be incredibly daunting at first, getting to grips with the economy. There are lots of tables and graphs to look out for, and sometimes things can be entirely out of your hands - if you end up with an economy that suddenly relies on Engine imports from Spain and Spain is suddenly attacked by its enemies, you might find yourself scuppered. Here's a quick primer to managing your internal markets.

Buy Orders And Sell Orders

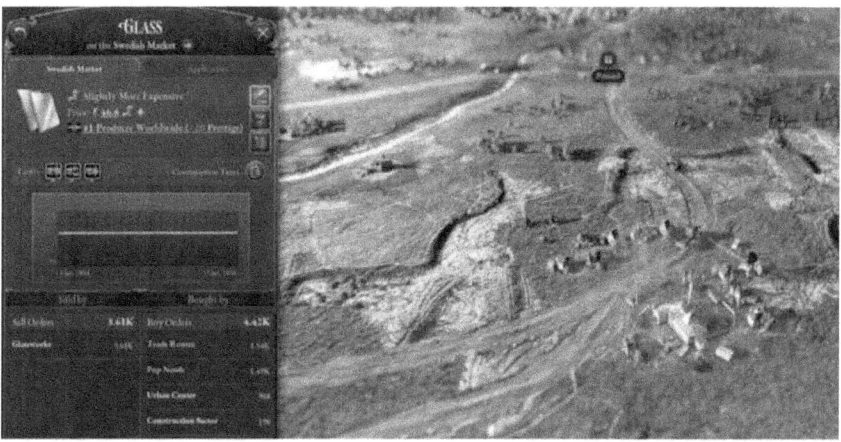

The simplest way to understand how a Good's price may be altered is through Buy Orders and Sell Orders - or, to put it more traditionally: supply and demand. All Goods in the game have a base price, which is an arbitrary figure that represents how much a person would pay for it if the country produced exactly enough of the Good to meet supply.

- If there's too much of a Good, its price drops. This means more people can access the Good, but buildings that produce the Good may experience difficulties.

- If there's not enough of a Good, its price rises. This means that the buildings producing the building can make lots of money and pay their workers higher wages, but those who wish to buy it are in a bad position.

In-game, you can check out a Good's Buy and Sell Orders in the Market menu.

- Sell Orders represent the Good's supply. This comes from your own production and any imports.

- Buy Orders represent the Good's demand. This comes from Pop consumption, building consumption, and exports.

If you want to increase the price of a good (for example, to make more money from a consumption tax or to fill your buildings' balances), you'll mostly want to be wary of oversupply. When expanding a Good-producing building, don't overexpand too much too soon - even if you have the raw materials to support it. This will lead to plummeting prices, which means the building won't be able to make enough money to pay its workers. This means less money for you through taxes.

Your GDP is calculated by taking the number of Goods you produce and multiplying it by the market price. If you want to increase your GDP a ton, you'll want to produce as much as you can while keeping the price nice and high.

Instead, you'll want to manage supply and demand carefully. If you start making too much of something, consider exporting it to other markets. Alternatively, alter the production methods in your buildings to produce less of it - even temporarily, this can make a huge difference.

Lowering the price of a Good is a bit simpler and is something you'll likely have to do quite often. You're bound to get notifications in the game telling you that certain Manufacturing Goods or Military Goods are too expensive - to alleviate this, you'll want more of them in your country. Either produce more of them by expanding your production buildings or upgrading your production methods, or import more of them from foreign markets.

The Types Of Goods

There are a few different types of Goods in the game, and this may alter your strategy for raising or lowering prices.

Good Type	Description	Examples
Staple Goods	These are the Goods that people need to survive. Without these, you can expect lots of radicalism. Having the price of Staple Goods swing either way is extremely bad - too expensive and your poorest Pops won't be able to survive, which means revolution, too cheap and your production lines will not be able to produce any, meaning even the wealthy won't have access to it.	Grain, Wood, Oil
Industrial Goods	These are goods that are used primarily in the manufacturing industries. They can be produced from Rural buildings depending	Iron, Dye, Glass

Good Type	Description	Examples
	on a state's resources, or they can be processed from those raw materials before being used to make other Goods.	
Luxury Goods	These are the Goods that people need to stay happy. They are harder to produce, more expensive to purchase, and are generally reserved for your wealthier Pops. Having these Goods be expensive is good for the economy in general, but you're relying on having the wealthy Pops to maintain an invested market.	Furniture, Tea, Art
Military Goods	These Goods supply military buildings, which are in-turn used to supply your armies.	Small Arms, Artillery, Ammunition

In general, you'll want your Staple Goods to be as affordable as possible without being unsustainable, your Luxury Goods to be as expensive as possible without being too detrimental to happiness, and your Industrial and Military Goods to be as balanced as possible to ensure your supply lines and army are sustained.

The Importance Of Infrastructure

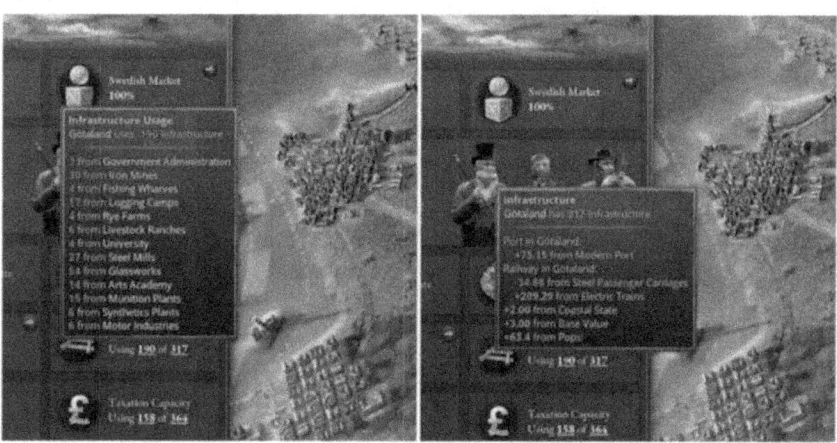

As your country grows, your infrastructure needs will grow alongside it. This makes sense - as you get larger cities and start supplying more Goods, you'd need more transportation to move them around your nation.

Buildings have infrastructure costs that increase as they are expanded, and states have infrastructure limits that depend on two different types of building: Ports and Railways. Ports contribute a little to infrastructure, but the majority will come from Railways.

If you start as a country that doesn't have Railways researched, you'll want to make it a priority as soon as it's realistic to do so.

If a state's infrastructure needs outweigh its infrastructure support, it will not be able to supply the goods it produces at an efficient rate. This means that they will sell for a lower price than the market price, regardless of what the market price actually is. It's very important to upgrade your infrastructure in line with your needs if you want to stay a stable economy.

VICTORIA 3: HOW TO RAISE YOUR STANDARD OF LIVING

From wages to welfare, here's how to improve your Pops' standards of living in Victoria 3

Most importantly, you don't want them to rise up, start a revolution, and kick you out of government or worse, start their own country. What you really, really want, is to make sure they have a good Standard of Living.

What Standard Of Living Affects

In Victoria 3, a Pop's Standard of Living is a measure of how able a person is to survive and then thrive. To put it into the simplest of terms:

- All Pops have a certain set of Needs. As a Pop's Wealth increases, their Needs will increase too, and they may end up needing a larger variety of Goods to satisfy their lifestyle. For example, Luxury Food Goods are only consumed by Pops of Wealth level 20 and above.

- Pops spend their wages on Needs.

- If a Pop has money left over after spending their wages on Needs, their level of Wealth will increase until their spending equalizes with their Need expenses.

There are three basic 'ranks' of occupation in Victoria 3, tied to how much they get paid by buildings:

- Lower Strata occupations - Clerks, Farmers, Laborers, Machinists, Peasants, Slaves, and Servicemen - are paid the least and require no qualifications.

- Middle Strata occupations - Academics, Bureaucrats, Clergymen, Engineers, Officers, and Shopkeeprs - are paid the average and require qualifications to acquire.

- Upper Strata occupations - Aristocrats and Capitalists - are paid the most.

Qualifications are gained over time, based mostly on Wealth and Literacy. For this reason, raising Literacy is a good way to enable people to move jobs a lot in order to acquire a better Standard of Living. Universities provide qualifications.

The Strata that a Pop belongs to will dictate their Expected Standard of Living - if their Standard of Living falls below this baseline, they are more likely to become unhappy and contribute to Radicalism and Turmoil.

As a result, a Pop's Standard of Living is basically equal to their Wealth level. If you want to raise a Pop's Standard of Living, you will need to make sure they can afford their Needs with enough left over to increase their Wealth. This will largely involve raising wages, but there are alternative strategies.

How To Raise Wages

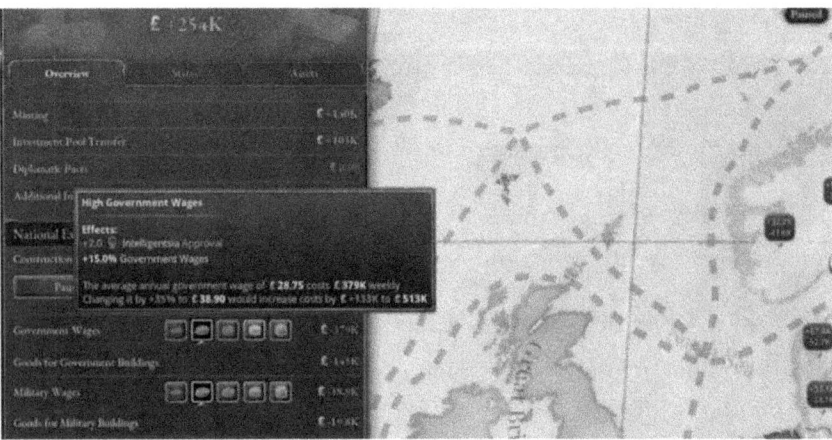

The best way to raise your country's Standards of Living is to make sure wages can rise. This involves making sure your buildings, which employ Pops, are profitable. Put simply, wages rise when a building needs to attract people to work there, and it can only do this when it has a positive balance.

You'll want to have a competitive workforce to ensure that wages stay high - don't put all of your efforts and development into few industries, and don't neglect any states in favor of, for example, your capital. While wages also depend on occupation, with lower paid jobs being an inevitability in the game, making an effort to raise all wages will improve your country's average Standard of Living.

If you Subsidize a building, it will be able to offer far higher wages. This draws from your country's treasury, but it's a great way to improve Standard of Living quickly and efficiently, as it's not beholden to market forces.

Government and Military employees have their wages dictated by you - you can alter these wages in the Budget menu. Increase them for a higher Standard of Living.

Having a healthy supply line is also key to satisfying Needs - you'll need to have plenty of Goods available, or prices will rise too high for Pops on lower wages to afford.

You can reduce taxes for an easy boost to Wealth - this means more money in Pops' pockets to spend on Needs.

How To Raise Welfare Spending

Another quick and easy way to raise your Standrd of Living is to introduce Welfare Spending. If enabled via your laws, Welfare Payments are given to Pops whose wages have them under the Normal Wage, which is decided through averaging many different wages across your nation.

Welfare is one of the laws you can change, with different levels that let you level up your Social Security institution. Higher levels of Welfare will make the threshold for receiving Welfare Payments lower, so more Pops will benefit from the extra money. This will lead to a drastic rise in Standards of Living at higher levels.

At the highest levels, the Social Security institution will severely limit the political power of those receiving payments - be wary of this.

In addition, enacting the Old Age Pension Welfare Law will increase Dependant Income and Dependant Enfranchisement, making your economy stronger at the cost of a small reduction to your Workforce Ratio. This is well-worth doing if your goal is a very high Standard of Living.

VICTORIA 3: HOW TO MANAGE THE MARKET

International trade is an important part of making money and supplying production in Victoria 3.

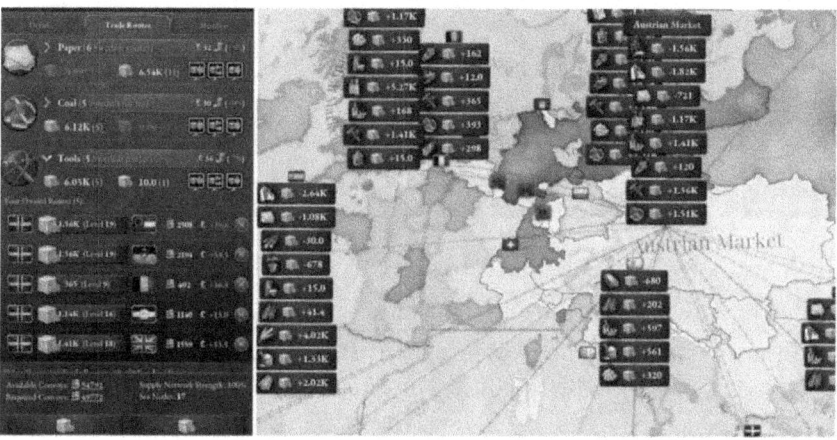

Unless you want to attempt a totally isolationist playthrough (which is possible, by the way!) of Victoria 3, you're going to have to learn how to deal with your national and international markets. Paradox Games are well-known for how complex their trade mechanics can be, but taking the time to delve into the system and learn how to use each moving part and turn them to your advantage is well-worth the effort.

With a bit of effort, you can dominate the international market, raise your Trade Centers to unfathomable heights, and command an army of Convoys that stretch around the world, importing and exporting everything from simple Grain to the finest of Fine Arts.

National Markets

In Victoria 3, every country has a national market unless it is the subject of another country, in which case it shares a national market with its overlord and any other subjects they may have. In a national market, Goods are attributed prices based on their total Buy Orders and Sell Orders, and this is one of the most important aspects of the game's economic simulation.

There are four different types of Goods in the game, but Victoria 3 treats them all in rather the same manner. They all have a base price, and their local price - that is, the price that your Pops will have to buy them at - is affected by the balance of Buy and Sell Orders.

In the Market menu, on the Members tab, you'll find a list of every state that takes part in the same national market as you. Alongside these, you'll find each state's level of market access - this is how well their infrastructure supports their production internally and externally. If a state has less than 100 percent market access, it'll be worse off when trying to purchase and sell Goods.

Imports And Exports

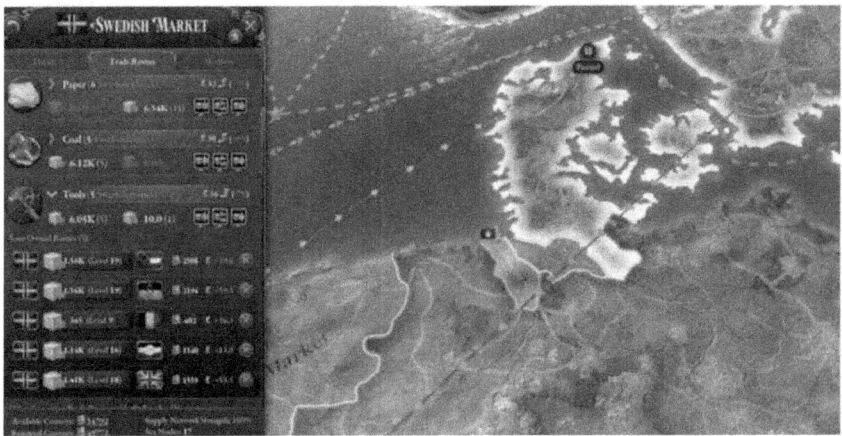

On the Trade Routes section of the menu, you can see an overview of all your current and potential trade routes - both exports and imports. For each Good, you can see a list of the routes you might want to open up, with details on how many units of the item you can expect to trade, how much trade revenue there's likely to be for the Trace Centers involved, and how the national market prices will be affected.

You can only trade with nations who are in the strategic regions that you have a declared interest in. It's important to note that you cannot declare an interest and then immediately trade with the nations there - it must be a sustained interest.

Tariffs And Market Goods Policies

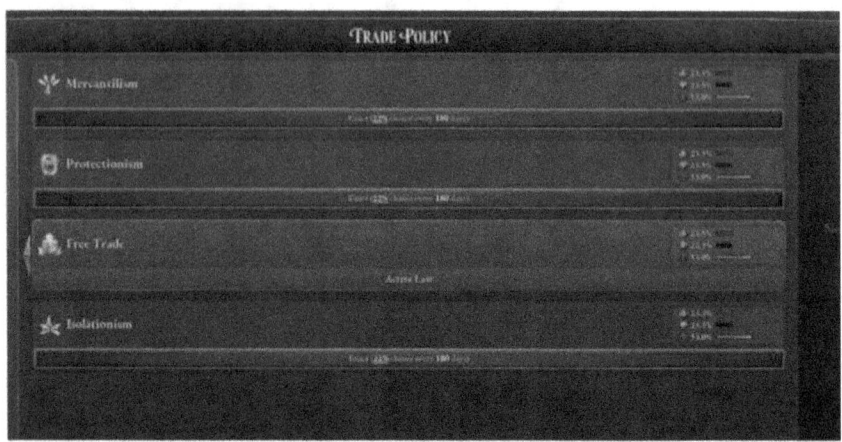

Tariffs are a tax on trade, affecting both imports and exports. When a Good is traded, it is expected that the revenue (or loss) is goes to the Trade Center building - the tariffs are deducted from this, going straight into the treasury. The amount of money brought in by a tariff is based on the tariff rate and the base price of the Good being traded, and tariff rate is based on two things:

- Trade Policy Law
- Market Goods Policy

Every nation has one of four different Trade Policy Laws, with two of them enabling the tariff system. If a nation has enacted either the Free Trade or the Isolationism Trade Policies, they do not have tariffs and therefore cannot set a Market Goods Policy for any Good.

When tariffs are in play, each Good being traded has a Market Goods Policy. By default, they are all set to No Priority, which gives a balanced leaning to imports and exports. There are two other settings, however - Protect Domestic Supply eliminates import tariffs to encourage more trade at the expense of your treasury, while Encourage Exports does the opposite, letting other countries buy up your Goods without having to pay tariffs.

Higher tariffs mean that your Trade Centers will be less likely to develop a trade route, as they will gain fewer bonuses to their wages and the Trade Center's revenue.

The table below details exactly how these Trade Policy Laws affect the Market Goods Policies.

Trade Policy Law	Effect on No Priority	Effect on Protect Domestic Supply	Effect on Encourage Exports
Protectionism	+10% tariffs on imports	+0% tariffs on imports	+20% tariffs on imports
	+10% tariffs on exports	+20% tariffs on exports	+0% tariffs on exports
Mercantilism	+15% tariffs on imports	+0% tariffs on imports	+30% tariffs on imports
	+5% tariffs on exports	+10% tariffs on exports	+0% tariffs on exports

Making trade decisions that encourage imports is good for an economy that's very heavy on production. This means the Protect Domestic Supply option is great for Goods that are used to make more expensive things, usually Luxury Goods, and good for getting more treasury out of export trade routes. The opposite is true for Encourage Exports - you can use this policy to offload a surplus in production to increase price a little, or gain more money from vital import routes.

Importing To Supplement Production And Needs

One aspect of trade that may be overlooked is how it can be used to great effect - and is sometimes entirely necessary - to supplement production. If you find yourself low on a resource that you require to make something, importing it is sometimes the way to go. If you're lucky, you'll be able to take advantage of cheap, prosperous trade routes.

When choosing a trade route to establish, the important numbers to look out for are the Amount and the Productivity. Amount is simply the number of units you'd get from the deal - you want this to supply your production needs, ideally. Productivity is how profitable the route will be, with the number being the wage increase that your Trade Center workers will receive.

You may find yourself in a situation where you need to import Goods just to supply production or your Pops' needs at all - this will come up from time to time with the rarer late-game Goods, such as Oil, Rubber, and Opium. The predictions made by the trade route menu here are pretty robust, so you can tell even at a glance if importing your entire supply of a Good will be profitable in the long run - you'll want to see the price of the Good in your national market decreasing in the tooltip when you hover over the prospective trade route.

Convoys

Unless you are a landlocked nation, you'll have to deal with Convoys sooner or later. Convoys are a resource produced by Ports. This resource represents your nation's ability to supply the trade routes that you have established and your ability to grow those trade routes into larger routes, sustaining more goods in the process.

If you are a trade-heavy nation, you will need a lot of Convoys, which involves building more, larger Ports and upgrading their production methods with bigger, more expensive ships. One problem you might run into at various stages of the game is that you run out of states in which to build Ports, which can be very annoying.

There are a few technologies to research that increase the maximum level of your Ports - these will be very important to get if trade is central to your economy. Get them as soon as you can.

VICTORIA 3: COMPLETE GUIDE TO WARFARE

The world is not a peaceful one in Victoria 3. Here's our guide to starting, managing, and winning wars.

While warfare isn't one of Victoria 3's main selling points, it's still a very important factor in the game. Most countries will end up at war sooner or later, whether it's to grow their empire or to defend their borders from antagonistic neighbors.

As you might expect, warfare is just as complex in Victoria 3 as many of its other mechanics. You have to manage your Generals and Battalions, enact laws that make your military stronger, and push warfronts with some strategy if you want to fight your way to success.

Diplomatic Plays - AKA, How To Start A War

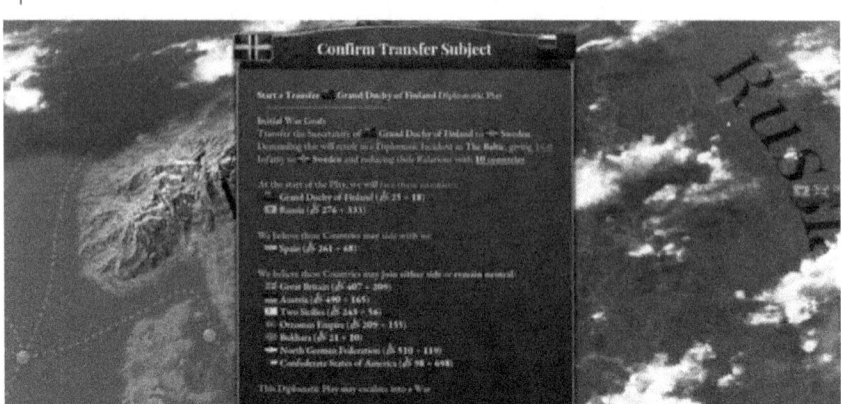

In Victoria 3, it is not possible to declare war outright. Instead, you must make a Diplomatic Play. Think of these as formal threats on a global stage in which you put your military might up against another nation, everyone with an interest in the region knows about it, and they can interfere at will.

The type of Diplomatic Play that you use to begin a conflict will affect what happens if your play is successful. Note that using different Diplomatic Plays will incur an infamy penalty, which will make the other countries with an interest in the same region like you less.

The table below details every Diplomatic Play and the outcomes they will cause:

Diplomatic Play	Effect
Annex Subject	The conquered vassal becomes a part of your country. You cannot annex puppets or dominions.
Ban Slavery	Slavery is forcibly banned in the target country.

Diplomatic Play	Effect
Conquer State	The states being targeted become a part of your country.
Cut Down To Size	Any states and subjects conquered by the target country over the last ten years are released.
Humiliate	Reduces the target country's prestige and prevents them from making Plays against you for a number of years.
Independence	Your country breaks free from its overlord.
Liberate Subject	You break the target free from its overlord.
Make Dominion	The target country becomes your dominion. A dominion is a subject that doesn't have to join its overlord's wars but pays them 10% of its weekly treasury income.
Make Puppet	The target country becomes your puppet. A puppet is a subject that has to join its overlord's wars and pays them 20% of its weekly treasury income.
Make Vassal	The target country becomes your vassal. A vassal is a subject that has to join its overlord's wars and pays them 15% of its weekly treasury income.
Open Market	Forcibly changes the target's trade laws to Free Market.
Regime Change	Forcibly changes the target's government and laws to match your own.
Return State	Returns the chosen states to your country - can only be used on states that consider your country their rightful home.

Diplomatic Play	Effect
Take Treaty Port	The chosen country cedes a Treaty Port to you - Treaty Ports open up trade between your countries and let you bypass tariffs.
Transfer Subject	Forces the target country to give you overlordship of the target subject.

That part about interest and regions is important, by the way - you can only make Diplomatic Plays in strategic regions in which you have declared an interest. The number of regions you can do this in depends on your prestige and power ranking, and some actions require that you have a sustained interest. In other words, it must have been declared for some time.

You won't go to war upon making a Diplomatic Play - instead, you'll enter the Escalation process.

Escalation And Maneuvers

Once you have chosen a Diplomatic Play, a countdown begins, and a new menu will be available on the right-hand side of the screen with a circle that fills over time. There are three stages to this process:

- Escalation, in which it is possible to add more war goals to the conflict. The war goals available will depend on the target country, with most of them matching the various Diplomatic Plays you could have made against them. War Reparations is a popular war goal that forces the country to pay you for a period of time after losing the war, but you can't make a Diplomatic Play for this alone.

- Diplomatic Maneuvers is the next, and longest phase. In this stage, it's time to sway other countries to your side. In the Sway tab, you can see where countries lean based on their position in the menu - those on your side are likely to join your side of the war even if you don't promise them anything. You can also see their army sizes here, showing you how valuable they can be in a war.

  - To sway a country, click on them in the Sway tab and then choose an option that has a green thumbs-up symbol next to it. Once chosen, you'll have to wait while they consider the offer. If they accept, they'll join your side in the war officially.

  - Offering Obligations is an easy way to get people on your side, but be wary of this. Obligations can backfire, as countries may oblige you to join in their wars.

  - It is possible to sway a country that has already pledged itself to your opponent's side. If accepted, they become neutral - they will not join your side.

- Once the Maneuvers stage is over, there is a short Countdown to War. At this stage, either side can Back Down - this will surrender the war goals in their entirety, but prevent the need for blood loss and economic destruction. AIs are likely to Back Down if you have an overwhelming advantage against them.

Adding war goals and attempting to sway countries to your side during this process costs Maneuvers. The number of Maneuvers you get to work with depends on your power rank and prestige. Be aware that both war goals and sway attempts use the same pool of Maneuvers, so adding lots of war goals will mean less chances to get more allies.

If you promise a war goal to an ally to get them to join the war, be aware that you'll have to make good on this in the peace treaty. If you don't there may be diplomatic repercussions.

An Introduction To Warfare

If neither country backs down during a Diplomatic Play, war begins. In Victoria 3, war isn't decided by moving stacks of soldiers - instead, there are moving warfronts to manage. Warfronts appear organically where the opposing countries meet, and both send their troops to clash there.

When one side is victorious at a warfront, the warfront gets pushed back further into the losing side's country, allowing the victor to make ground and win warscore. Winning battles and taking land will improve your country's standing in the war, and your opponent will lose standing, becoming more willing to accept a peace deal or capitulate outright.

Generals And Battalions

To actually fight in a war, you will need two things: Generals and Battalions.

Generals are recruited in the Military menu and are necessary to lead your soldiers into battle. If a soldier is garrisoned and not assigned to a General, they simply do not get to participate in a war.

- Hiring and promoting Generals costs a certain amount of Bureaucracy, so be wary of your balance before starting or escalating a conflict.

- Generals can only support and lead a certain number of troops - if you want them to lead more, you'll need to Promote them. To do this, click on their portrait in the Military menu to open up their detailed menu and Promote them from there.

    - Be wary that Promoting a General will increase the political power of any Interest Groups that they are a part of.

Generals all have traits - taking a look at these traits may help you decide where to mobilize them and which tactics to use. For example, a General with Expert Offensive Planner will obviously have greater success when using offensive tactics and Advancing Fronts.

Battalions are recruited at the Barracks building, with one Battalion arising from each fully-staffed level of Barracks. The production methods used in your Barracks will affect the quality of your soldiers, so be sure they're well-supplied and up to date when you start a war.

If you have laws that enable Conscription, you will also have access to Conscription Battalions during war. These must be enabled individually in the Military menu, with Conscripts joining the forces of any available Generals.

Conscription Centers pop up during Diplomatic Plays - this is when you should change up their production methods to get better results from Conscripts.

Battalions and Generals are location-based - you will only be able to mobilize Battalions in strategic locations once you've recruited a General there. For this reason, it's a good idea to consolidate your military development into a few, select regions - this is only relevant if you have quite a large country, of course.

The Frontline

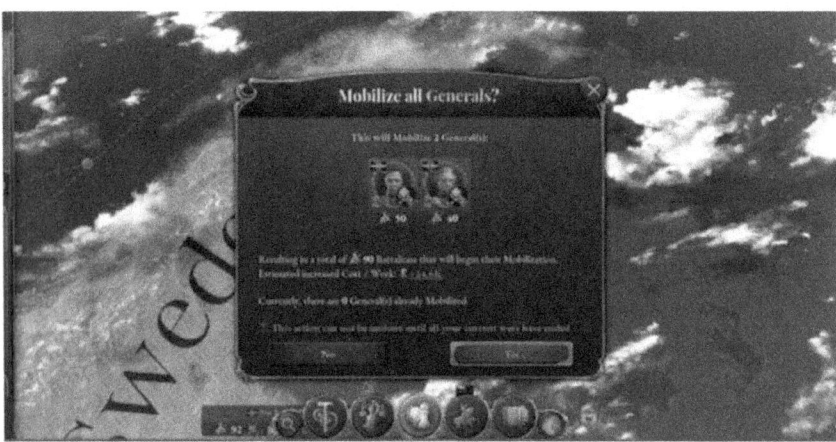

To see which frontlines are currently active, you can either look for them on the map or click on the War Overview menu on the right-hand side of the screen. This will give you a quick look at how many Battalions you have stationed there and what your general chances of victory at that front are.

You will need to mobilize your troops when war breaks out. Do this from the Military menu and pick the warfront that you want each General to move towards - note that it will take a period of time before they arrive, so do this sooner rather than later. You can mobilize Generals before war breaks out, during Diplomatic Maneuvers.

After a certain amount of time has elapsed at a warfront, battles will start happening. This will pit Generals against each other, with their Battalions doing all the fighting. You can watch the battle's progress by clicking on it in the Warfront Overview menu.

Naval Warfare

Similarly to Generals and Battalions, coastal countries have access to Admirals and Flotillas. Admirals are recruits just like Generals and have the same qualities and limitations, while Flotillas are recruited from Naval Bases in the same way that Battalions come from Barracks.

Rather than heading to fronts, Admirals and their Flotillas can perform a few different maneuvers during war.

- Patrol Coast will have them do exactly that - if they meet an enemy Admiral, they will fight in the same way that land-based armies do.

- Raid Convoys lets you select a hostile supply network to raid. If successful, it will destroy some of the enemy's convoys, and therefore trade.

- Escort Convoys lets you protect convoys in one of your supply networks, in an attempt to disrupt enemy attempts to raid.

- Naval Invasions let Admirals escort Generals and their Battalions to an amphibious landing. This can be useful for circling around a coastal country and taking them from two sides, and is necessary when starting an overseas war as there won't be any frontlines to fight without them.

    - You cannot conduct a Naval Invasion without a General. If you choose a General to lead a Naval Invasion that is already fighting at a front, they will be pulled from the front to conduct the invasion.

Peace Deals

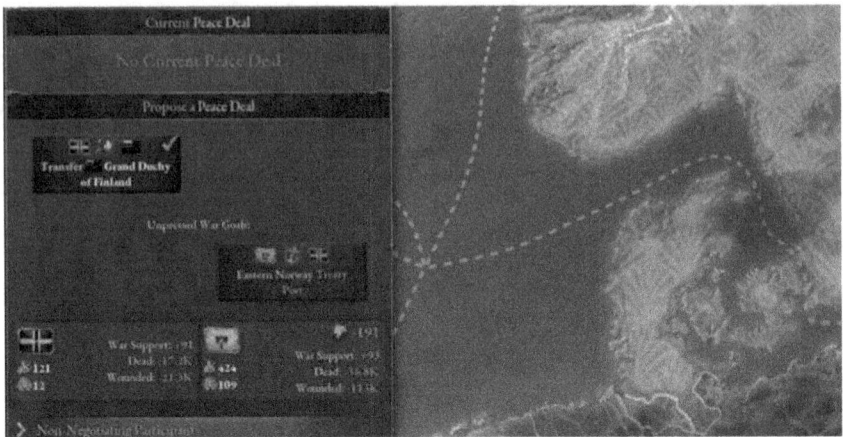

There are three ways to end a war:

- Propose a successful peace deal or accept one from your foe.

- Propose a successful white peace or accept one from your foe.

- Capitulate, or have your enemy capitulate.

To propose a peace deal, click on the War Overview and then head to the Make Peace tab. Here, you'll find the applicable war goals. Click on any unpressed war goals to press them, which sets the details of your proposed peace deal.

Note that you can agree to some or all of your foe's war goals, as long as they don't directly interfere with your own. This will make them more likely to accept the peace deal, as you can tell from their acceptance level next to their flag.

A white peace is simply a peace deal with no war goals pressed. For an AI to accept this, the war has to be devastating enough to make them really want an end to it.

Capitulation means accepting every war goal against you, including the unpressed ones. A sufficiently devastated enemy AI player will capitulate, ending the war immediately.

When a peace deal is reached, an information box will pop up showing you every war goal that was in play. Only the war goals with check marks next to them have been resolved, though!

Warfare Tips - AKA, How To Win A War

- There's not really a lot of micromanagement required in Victoria 3 - success in war will require more forward planning than fast reactions.

  - Having allies and a good diplomatic standing will allow you to secure more allies during the opening phases of a Diplomatic Play.

- That said, don't rely on AI allies to win wars for you. Always consider that they may have their own problems preventing them from mobilizing their entire army, or they may have other problems. Their huge armies may not be updated, for example.

- While army quality is important, having huge numbers of soldiers is always a good shout. Consider building lots of Barracks before a large war and enacting laws that let you mass-conscript your civilians. If a war is that important, it'll be worth the short-term deficits in money and happiness.

- If fighting a war against a large coastal country, use naval invasions to force them to fight on multiple fronts. If you have the numbers advantage, pushing one front massively while distracting much of their army with the other is a great way to gain ground quickly.

- Events will pop up during wars a lot, often forcing you to choose between different negatives. We recommend avoiding troop recovery maluses - troop recovery is very important in protracted wars.

- Don't neglect the running of your country during a war. You'll likely have trade to sort out, radicalism to keep an eye on, and supply line issues to deal with. If you let these fall to the wayside, you may start losing the money that you need to run the war.

VICTORIA 3: HOW TO MANAGE YOUR INTEREST GROUPS

Understanding the ins and outs of Interest Groups will let you mold your country to your whims and desires far easier in Victoria 3.

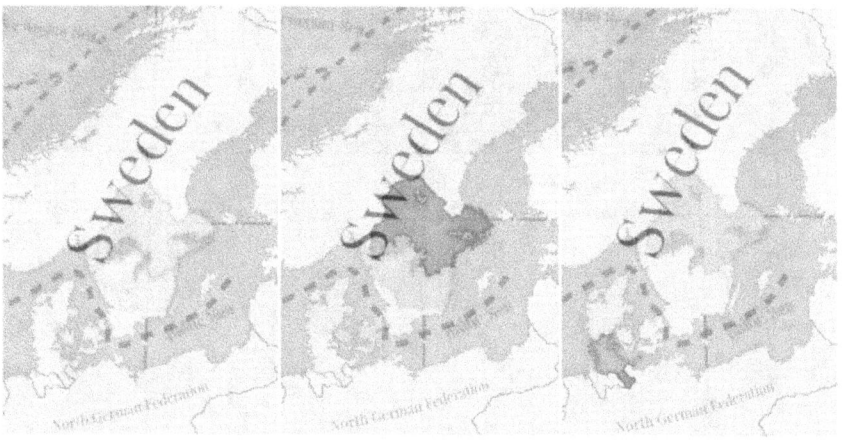

Victoria 3 features a very in-depth political simulation and managing it is one of the biggest parts of the game. You have the ability to suggest new laws, for example - but doing so will require the approval of your people. Specifically, this means your Interest Groups.

If you go in with all guns blazing, attempting to change the very essence of your country's governance without an ounce of gradation, your people will rise up against you. Instead, gradually changing the makeup of your government and using both obvious and subtle techniques to manipulate those with political clout is the key to success.

Interest Groups And How They Form

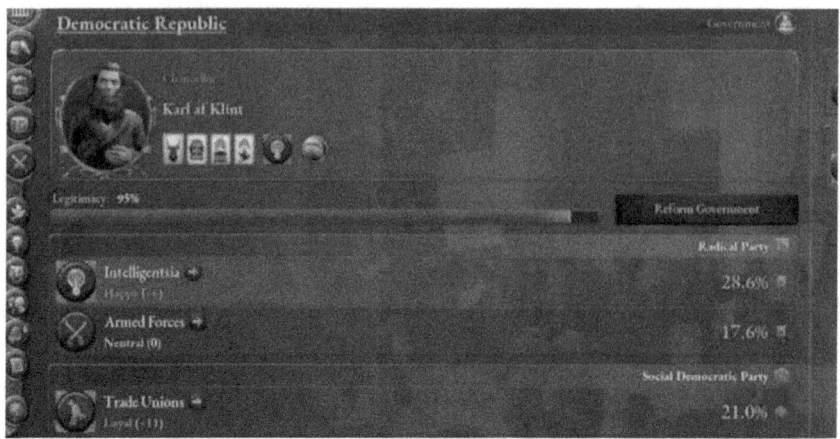

Interest Groups are politically-motivated groups of people who band together thanks to a common interest, which will dictate what kind of politics they favor. For example, the Rural Folk will be far more in favor of lower taxes for their income group and a focus on developing agriculture, while the Intelligentsia desires more academia and less religion.

Interest Groups get more Pops to join them through attraction, which is based on factors like wealth, certain professions, and education access. Occupation is one of the biggest driving forces, with, for example, laborers being far more likely to join a Trade Unions group if one exists.

If your country has voting rights (i.e. is not Autocratic, Oligarchic, or Ararchistic), Interest Groups may form together to form political parties. This means they'll be grouped together when considering whether or not they can be made part of the government or not, and they cannot be separated. The Interest Groups within the party are still separate, but are empowered by the votes that the party earns in an election based on their Clout levels.

All Interest Groups have leaders, and leaders have traits that can have substantial effects. Notably, every leader has an Ideaology which will alter the entire Interest Group's Ideaologies - more on that later. Depending on your government type, your nation's leader may end up being the leader of the Interest Group or political party with the most Clout. This is not an issue with government types that have a singular leader chosen differently though, such as a Monarchy.

Ideaologies And Traits

Interest Groups have two features that you need to be aware of. The first is Ideologies: these are a number of belief systems that dictate which laws the Interest Group approves and disapproves of. Each individual Idealogy has a number of stances that they take - for example, the Anti-Clerical Ideology takes stances on the Church and State, Education System, and Bureaucracy laws.

Not every Interest Group with the same name will have the same Ideaologies as a rule, and unless their leader is a Moderate, every Interest Group will have an extra Ideaology (listed at the top with a different color icon) that matches the leader's.

Interest Groups also have three Traits - two positive, and one negative. These traits become activated depending on their approval, which is dictated by your current laws and any short-term modifiers that can come about thanks to events.

- If the group's approval is five or above, the first positive Trait will be activated.

- If the group's approval is ten or above, both positive Traits will be activated.

- If the group's approval drops to minus five or below, the group's negative Trait will be activated.

Negative Traits can be highly damaging, while positive Traits can be fantastic boons for your country, so trying to appease every Interest Group that isn't Marginalized is not a bad idea.

Approval

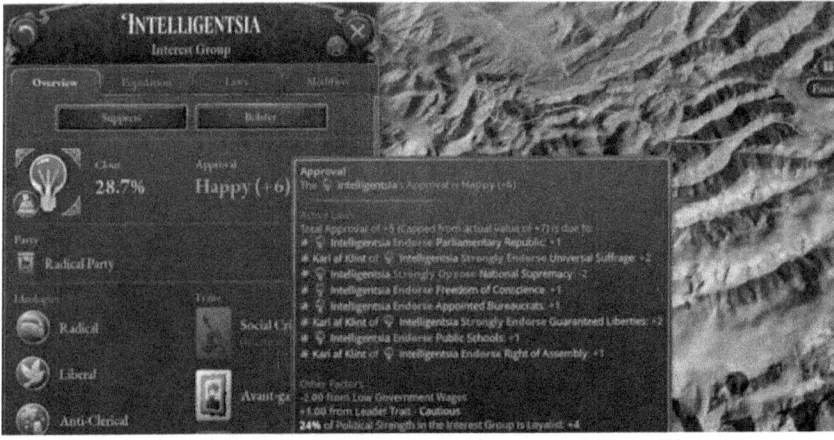

Every Interest Group has an approval rating that shows how much they are in favor of the currently enacted laws. The laws that affect a group's approval depend entirely on the laws that they take a stance on thanks to their Ideaologies.

Also affecting approval is the balance of Radicalism to Loyalism within the Interest Group, moderated by political strength. This means that if there are lots of members of the Interest Group with high political strength and they are radical thanks to being unhappy with something, the Interest Group will have a lower approval level.

Approval can also be affected by many disparate factors, such as certain taxes, random events, individual leader traits, and currently-debated laws.

An easy way to see which laws would boost a group's approval is to click on the Laws tab of the Interest Group's overview menu. This will present you with a list of laws that are currently available to enact that would increase your approval.

Clout

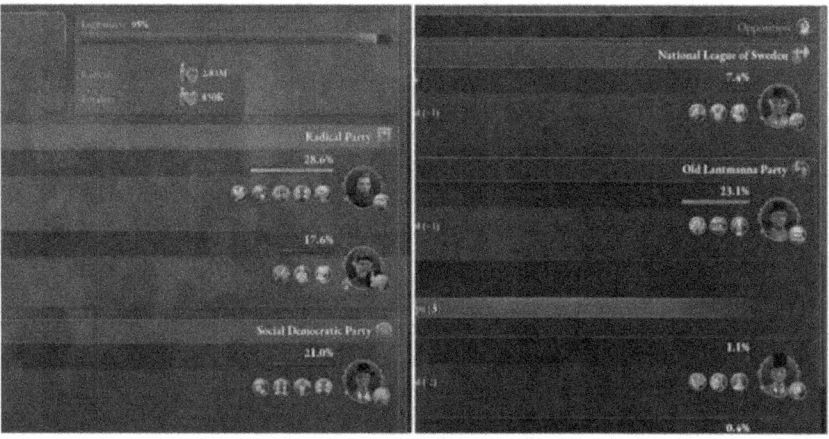

Clout is a rating of how politically powerful an Interest Group is. The more a group's Clout, the more it's able to affect the governance and lawmaking processes for your country. Your government's Legitimacy is highly dependent on Clout - if your government isn't made up of Interest Groups with Clout (read: political power), it won't be considered legitimate, and you might have more radicalism to deal with as a result.

Wealth plays a huge role in determining Clout. The richer a Pop is, the more power they contribute to the Interest Groups they have a stake in, if any. This information provides you with the best means to increase or decrease a group's Cloud - enrich those with membership. This can be done by increasing the wages and the job opportunities for the occupations strongly represented in those groups.

Pops in a country's capital will contribute more political power to a group's Clout than Pops outside the capital. Conversely, Pops in unincorporated states contribute far less political power.

For example, the Trade Unions Interest Group is likely to have lots of Machinists, Clerks, and Laborers. To make this group more powerful, you can improve their working conditions and invest into their places of work, which are likely to be Industrial in nature. Conversely, Capitalists are likely to be part of the Industrialists group, and could be empowered by changing buildings' ownership to Publicly Traded.

There are three basic 'levels' of Clout that determine an Interest Group's overall relevance and power:

- Above 20 percent, an Interest Group is considered Powerful. These groups' Traits become twice as effective as a result. This includes their negative traits so it's important to keep them happy.

- Between five and 19 percent, an Interest Group is considered Influential.

- Under five percent, an Interest Group is considered Marginalized. This means their Traits are never active, regardless of their approval level. An Interest Group in power cannot become Marginalized.

Bolstering And Suppression

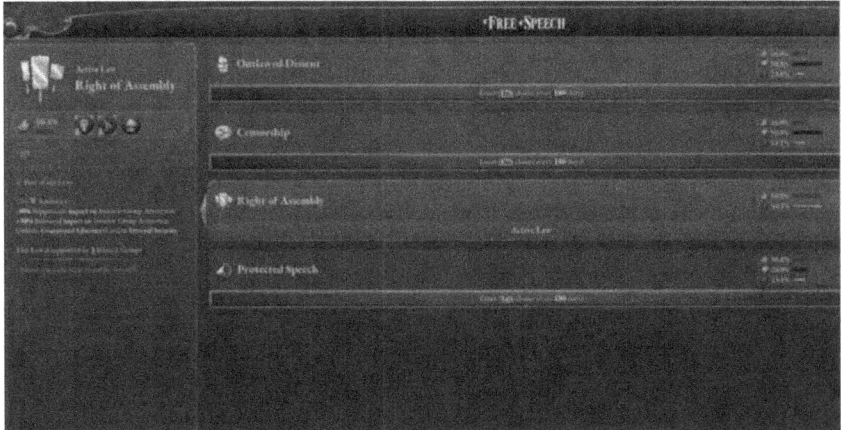

One of the ways you can directly interact with an Interest Group is through the Bolster and Suppression actions. These can be taken from the Interest Group overview menu.

Bolstering an Interest Group increases its attraction. This means more Pops will join the group, giving it more Clout.

Suppressing an Interest Group decreases its attraction. This means fewer Pops will join the group and current members may wish to leave. You cannot Suppress a party that is in power.

Using either of these actions costs a good deal of Authority, so it's a good idea to keep an eye on the group's Clout and stop Bolstering or Suppressing once the desired level has been reached.

The efficacy of your Bolster and Suppression actions depends on your Free Speech laws. The harsher the laws, the more effective they'll be, and if you enact Protected Speech, you won't be able to use these actions at all.

ABOUT THE AUTHOR

When I finding new tricks, tips, and strategies to beat each other, they came up with a brilliant idea. Let's take these hours of gaming expertise, and share these skills with like mind people. At that moment, the Victoria 3 Complete Guide were born. With more exciting gaming books being developed in the Lab as we speak. I am creating a buzz in the gaming guide publishing world, with a ground swell of followers, anxiously awaiting my new releases.

www.ingramcontent.com/pod-product-compliance
Lightning Source LLC
Chambersburg PA
CBHW071127240526
45465CB00024B/1437